de Gruyter Studies in Organization 29

The Management of Technological Learning

de Gruyter Studies in Organization

Innovation, Technology, and Organization

A new international and interdisciplinary book series from de Gruyter presenting comprehensive research on the inter-relationship between organization and innovations, both technical and social.
It covers applied topics such as the organization of:
- R & D
- new product and process innovations
- social innovations, such as the development of new forms of work organization and corporate governance structure

and addresses topics of more general interest such as:
- the impact of technical change and other innovations on forms of organization at micro and macro levels
- the development of technologies and other innovations under different organizational conditions at the levels both of the firm and the economy.

The series is designed to stimulate and encourage the exchange of ideas between academic researchers, practitioners, and policy makers, though not all volumes address policy- or practitioner-oriented issues.
The volumes present conceptual schema as well as empirical studies and are of interest to students of business policy and organizational behaviour, to economists and sociologists, and to managers and administrators at firm and national level.

Mark Dodgson

The Management of Technological Learning

Lessons from a Biotechnology Company

Walter de Gruyter · Berlin · New York 1991

Dr. Mark Dodgson
Science Policy Research Unit, University of Sussex, Falmer, Brighton, Sussex, U.K.

⊗ Printed on acid-free paper which falls within the guidelines of the ANSI to ensure permanence and durability.

Library of Congress Cataloging-in-Publication Data

Dodgson, Mark, 1957–
The management of technological learning: lessons from a biotechnology company / Mark Dodgson.
p. cm.--(De Gruyter studies in organization ; 29)
Includes bibliographical references and index.
ISBN 0-89925-768-2 (U.S.)
1. Celltech Ltd.--Management. 2. Biotechnology industries--Great Britain--Management--Case studies. 3. Technological innovations--Great Britain--Employee participation--Case studies. I. Title. II. Series.
HD9999.B444C453 1991 90-25326
660'.6'068--dc20 CIP

Deutsche Bibliothek Cataloging-in-Publication Data

Dodgson, Mark:
The management of technological learning: lessons from a biotechnology company / Mark Dodgson. – Berlin ; New York : de Gruyter, 1991
(De Gruyter studies in organization ; 29 : Innovation, technology, and organization)
ISBN 3-11-012706-7
NE: GT

Typesetting: Knipp Textverarbeitungen, Wetter – Printing: Gerike GmbH, Berlin – Binding: Lüderitz & Bauer, Berlin – Cover Design: Johannes Rother, Berlin.

For Yo

Preface

Why is it that rapid technological change is a threat to some firms and an opportunity for others? This question is of fundamental interest to managers, academics and public policy makers concerned with the technological basis of competitiveness. It is the intention of this book to describe and analyse a key feature of the processes which allow technology to be used by firms to assist their competitive position. It is concerned with the ability of the firm to learn.

The book argues that the differential ability of firms to learn affects not only the competitiveness of individual firms, but can also direct the development of new technologies and industries. This argument is made in the context of the development of biotechnology, in circumstances where rapid and extensive change places a premium on learning capacities. Biotechnology provides an excellent example of the ways in which the pressures to commercialise new scientific discoveries highlight the advantages accruing to firms adept at dealing with novelty. One of the major actors in this commercialisation process is a new type of firm; the small, dedicated biotechnology company (DBF). The existence and future contribution of such firms depends on their ability to learn quickly.

The DBF, and the complex and changing nature of the relationships between academia, DBFs and large established firms, are major factors which have stimulated and directed the development of biotechnology. Attempting to analyse the emergence of a new technology, and its scientific and commercial development, is a phenomenally complex task. It is necessary to do so as it is only by understanding the past and current pressures on emerging technologies that one can attempt to influence their future development. Through an examination of the role played by DBFs – and one DBF in particular – and their competitive and collaborative relationships, the book attempts in a small way to contribute to the understanding of some of the factors which have affected and will affect the scientific and commercial progress of biotechnology. It aims to provide one piece of the jigsaw that is a complete picture of the evolution of a new technology.

The book reports on the development of an important DBF, Celltech Ltd. It provides one of the very rare examples of a study of how a firm actually accumulates technological know-how in order to establish a competitive position and to grow. It details the long-term learning processes by which a company builds its core technology base. A number of strategic management of technology issues feature, including: the management and organisation of Research and Development and technological collaboration; matching technology strategy with human resource development strategy; and engendering a creative organisation and culture.

The high significance of DBFs in the emergence of biotechnology has attracted considerable attention from those concerned with explaining the technology's pattern of development. Hall (1987) describes the initial commercialisation of the science of biotechnology and focusses on the emergence of two of the earliest DBFs: Genentech and Biogen. Teitelman (1989) analyses the problems of financing its development, and takes to task much of the hype over biotechnology, by referring to another DBF: Genetic Systems. The case study of Celltech is used to refer to a broad range of issues, including scientific and financial, but it concentrates particularly on organisational and managerial concerns.

It will be argued that DBFs' past and future contribution to biotechnology's development and diffusion rests with strategies for fast technological learning: the ability to acquire, develop and operationalise technological know-how quicker than their competitors. It is this which potentially gives DBFs a comparative advantage over large multinational companies, and has been a major stimulus to high level of interaction between them and large established firms. The ability of DBFs to establish creative, well organised and highly motivated Research and Development (R&D) departments, to work effectively with the science base, and to benefit from collaboration with larger industrial partners are all features of technological learning, and are crucial to DBFs' competitiveness and the role they play in the propagation of the technology. It will be shown how developing and implementing such strategies requires considerable attention be placed on management and organisational issues. In consequence, these issues which are particular to individual firms, affect the development of a new technology as a whole.

The use of industrial case studies to illustrate complex situations is not uncommon in some of the social sciences. Ron Dore's (1973) comparison of British and Japanese industrial relations was based essentially on two case studies. Some of the richest industrial sociology has been based on studies of single firms, such as Beynon's (1973) study of Ford, Beynon and Nichol's (1977) study of 'Chemco', and Roy's (1958) and Burawoy's (1979) study of 'Allied'. The case studies are used as they are particularly revealing about much broader issues.

The book is structured in a format which mixes description with analysis. Chapter 1 is concerned with the impact of biotechnology on industry, and incorporates both theoretical and practical considerations. Chapter 2 describes Celltech in 1990. Chapter 3 describes its genesis, and illustrates subsequent chapters on how the company's activities and achievements are in many ways a legacy of its past. Chapters 4 and 5 are descriptive, intending to depict how companies grow. Chapter 6 is both descriptive and analytical and is concerned with the ways Celltech built a creative, learning organisation. Chapters 7 and 8 are also both descriptive and analytical and are concerned with how Celltech has build a leading technological capability on the basis of its internal and external learning activities. Chapter 9 addresses learning more systematically. It considers the link between strategy and learning and conceptualises the role of learning in competitiveness and corporate and technological development.

Many people have assisted my learning processes in the production of this book. I was extremely fortunate to be asked to write the history of Celltech, and learnt a great deal during the year I spent doing so. I am grateful to the Economic and Social Research Council's Designated Research Centre in Science, Technology and Energy Policy at the Science Policy Research Unit for funding the study and subsequent period when the book was written. Celltech Ltd and the British Technology Group generously contributed to the funding of the writing of the history. The study of Celltech is published as a SPRU Special Report (Dodgson, 1990a), and provides the basis for this book.

Many very busy people within Celltech were kind enough to give me their time during the research. I am particularly grateful to Gerard Fairtlough, John Jackson, Gwyn Humphreys and Geoff Brooker in this regard. My colleagues at SPRU have provided me with constant support and good advice. I am particularly grateful to Roy Rothwell and Keith Pavitt. Margaret Sharp's encouragement, insights and detailed comments have been immeasurably valuable. Thanks are also extended to Chris Freeman, Smail Ait el Hadj, Martin Bell and Aubrey Jones. Any shortcomings in the book exist despite my colleagues' attempts to prevent them.

Every company makes mistakes and endures often painful failures. Celltech is no exception. It has had its share of problems. Nevertheless, one of the major lessons I have learnt during the last two years studying and writing about Celltech is what can be achieved with vision and energy. These underpin the company's achievements, and many people inside and outside of the company have combined them to create Celltech. The two virtues are personified in Gerard Fairtlough, who has had the vision to see the necessity for new forms of organisation, and the energy to put his ideas into practice. Whatever the future holds for the company – and it es operating in very uncertain circumstances – its achievements so far are enough to prove that the imagination and creativity of Britain's scientists can be matched in the management of the commercial development of that science. This in itself is an important lesson.

Mark Dodgson July, 1990

Contents

List of Figures and Tables

Chapter 1
Biotechnology and Industry

Introduction

Biotechnology is one of the most important technologies to have emerged over the past twenty years. Along with information technology and new materials technology it is believed to be one of the 'generic' technologies which will underpin much future industrial growth. It is a radically new technology, capable of profoundly affecting existing ways of doing things, and has the potential to pervade a number of important industrial sectors. Biotechnology's novelty and scope inevitably causes considerable turbulence and change within firms, and in the relationships between firms using it. It has provided the basis for the development of a new type of firm: the dedicated biotechnology firm (DBF). The rapid changes occurring as a result of biotechnology provides the context in which the study reported in this book is placed. One of the cardinal questions posed is: how do industrial firms deal with comprehensive technological change?

The Emergence of Biotechnology in the Science Base [1]

Throughout this book the term biotechnology will refer to the 'new biotechnology' that has emerged in the past 15 to 20 years. It includes the use of recombinant DNA and cell fusion techniques, and bioprocessing technology, to make or modify products. The products targeted lie within three broad sectors: health care (therapeutics and diagnostics); agriculture; and waste management; although it also has applications in chemical biotransformations, energy and mineral recovery. Health care is currently the centre of corporate attention in biotechnology and will be the focus of this book. [2]

Two separate scientific discoveries formed the basis of the 'new biotechnology'. These were reported in the early 1970s in two research laboratories; one British, one American. Milstein and Kohler at the Laboratory of Molecular Biology (LMB), Cambridge reported the discovery of monoclonal antibodies (MAbs). Boyer and Cohen at Stanford University reported the discovery of recombinant DNA (rDNA). MAbs, or hybridoma technology, involves fusing together cells with specific properties to create a method of producing large quantities of specific antibodies. Recombinant DNA, or genetic engineering, involves splicing a 'foreign' component into the genetic structure of an organism so as to change its characteristics or functions – eg to make it produce insulin or other useful proteins.

Although the science of these two process innovations was characterised by very considerable uncertainty and speculation, and expectations of actual products were unclear, biotechnology's potential excited high levels of interest from many quarters. To some extent expectations of it were measured against what had happened in information technology. The computer industry and IT provided a model whereby a new technology had rapidly created huge new markets and a number of very large new firms, as well as making some technologists and financiers very rich. Almost immediately the race was on to develop and commercialise the two scientific discoveries.

At this time the new skills of the biotechnologists resided almost entirely amongst scientists in universities and hospital schools. Following the model in electronics, many of these scientists in the US – molecular biologists, biochemists, microbiologists, and molecular geneticists – decided either to form or work closely with commercial organisations primarily in order to benefit financially from their scientific know-how. Since biotechnology's early commercial development depended critically on the transfer of these skills from the science base, [3] the DBF proved one of the most important mechanisms by which the links between the science base and industry were established.

The Dedicated Biotechnology Firm

DBFs began to emerge in the United States in the early to mid-1970s. It is estimated that there are now over 400 such firms in the USA, and over 80 in Europe (Office of Technology Assessment (OTA), 1988:5 and Orsenigo, 1989:152). The largest DBF, Genentech in the USA, 60 per cent of which was recently acquired by a large European drugs company for $2.1 billion, has sales in excess of $300 million and employs over 1750 people. Figure 1 shows the growth in the number of DBFs in the USA and Britain. It shows how since the early 1980s the numbers of DBFs being formed has declined. Indeed, a recent report to the US Congress relates how some analysts consider the number of viable DBFs to be declining (OTA, 1988:12). The reasons for this decline will be examined in later sections. As shown in Table 1, some leading DBFs have enjoyed rapid patterns of growth.

The emergence of DBFs in the USA has been attributed to a number of interrelated factors. Important amongst were: the development of the novel scientific techniques in universities and research hospitals; the ready availability

Table 1. Turnover in Some Leading DBFs (1988/89)

Company	Turnover £ Million
Genentech	118.1
Cetus	36.0
Centocor	36.8
Chiron (89)	20.0
Amgen (89)	46.8
Celltech (89)	19.7
Transgene	5.5

of venture capital; the entrepreneurial drive of scientists, venture capitalists and

Figure 1

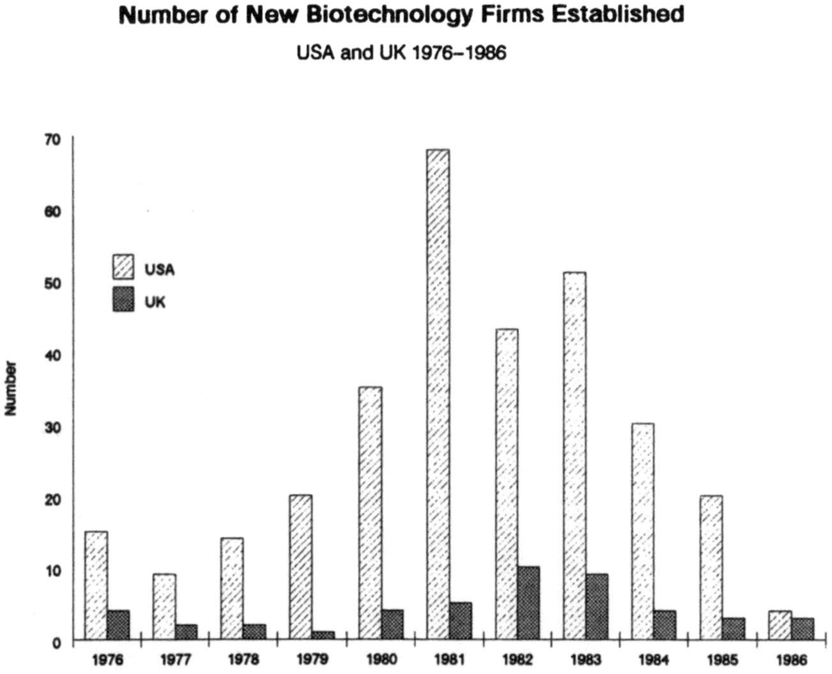

Source: OTA, 1988, and Oakey et al., 1990

intermediaries between the two; and the synergistic relationships DBFs formed with large firms.

Hall's (1987) account of the formation of Genentech, a company initially created by an alliance between a venture capitalist and some entrepreneurial scientists, highlights some of the factors which led scientists to work for or with DBFs. Important, of course, was the possibility of large financial rewards through the high salaries and stock options which a fast growth company could provide. But also important were the possibilities provided of being able to undertake well-funded research not dissimilar to the research undertaken in academia, and avoiding the vagaries of decision-making by public grant-giving bodies. The research could be undertaken, moreover, in an informal environment very similar to that of a university and unlike the bureaucratic arrangements often found in large established firms.

Teitelman (1989) describes Wall Street's initial attitude to biotechnology as 'biomania'. In 1980/81 biotechnology investments in the USA were attracting nearly $100 million of venture capital (Hacking, 1986). Wall Street's relationship with biotechnology has been extremely variable, and it has blown hot and cold on a number of occasions. However in its early enthusiasm for the technology it ensured that many DBFs enjoyed substantial funding.

The relationships between DBFs and large, multinational companies are one of the major features of the development of biotechnology, and will provide one of the focuses of this book. Large, established pharmaceutical/chemical firms were generally slow to become involved in biotechnology, but currently are devoting considerable resources to it, and many have acquired DBFs. Some large firms were involved in biotechnology very early – for example ICI in Britain, and Eli Lilly in the USA. But few risked major investments and until the early and mid-1980s the majority of industrial activity in biotechnology lay in DBFs, with large, established firms tending not to invest heavily, but keeping a 'watching brief' on DBFs' efforts. Indeed, the 1988 OTA report (p.80) describes a survey which concluded that as late as 1987, DBFs played the major role in industrial biotechnology R&D. DBFs spent $1.2 billion on R&D compared to large corporations' $0.8 billion. However, much of the DBF's expenditure derives from contracts with large firms. The report also suggests that major corporations' expenditure is increasing comparatively more rapidly.

There are a number of other reasons why it was DBFs rather than large firms which initially developed biotechnology. One reason lay with the unproven and risky nature of the technology itself. If DBFs were developing the technology, and undertaking the risk, why should large firms bother developing their own competences when they could easily acquire them when the science was better established, and market opportunities clearer? Conventional pharmaceutical companies' R&D structures tended to be arranged in line with the disciplines of biology and chemistry. The new skills of molecular biology did not accord well with these existing structures. Furthermore, the traditional methods used in the search for new drugs – the search for new substances, then extensive screening for potential applications – were very different from the possibilities provided by biotechnology whereby drugs could be 'designed' with a particular application in mind. As Dosi (1988) puts it, "What the firm can do technologically in the future is narrowly constrained by what it has been capable of doing in the past". Another reason for the important role of DBFs in the early development of biotechnology lies with the focus of this book, namely the differing nature of the organisation and management of DBFs compared with that of large firms. Radical new technologies like biotechnology, with the potential profoundly to alter existing structures and systems, put a premium on firms with the ability to learn rapidly about new scientific and market possibilities, and the personnel and organisational structures to take quick advantage of these lessons for commercial benefit. It was the DBF which, being new, had the opportunity to create novel management systems and practices and organisational arrangements in order to take advantage of the opportunities biotechnology provided.

The Relationships Between the Science Base, DBFs and Large Firms

One analysis of the relationship between academia, DBFs and established firms is based on the work of David Teece. Teece suggests that for firms to realise the full return from their investment in creating a technological innovation they need to access what he calls 'complementary assets' (Teece, 1987). In biotechnology such assets might include: competitive manufacturing; marketing and distribution networks; and the ability to deal with the regulatory procedures involved in getting new products onto the market. It is argued that the academics, DBFs and large firms have each brought distinctive advantages to the creation of biotechnology innovations (Pisano, Teece and Shan, 1988). Initially, scientific expertise lay in the science base, and was gradually transferred to DBFs as they developed their own R&D competences. Large firms initially lacked the R&D competences to innovate in biotechnology themselves, but they had all the complementary assets necessary to commercialise discoveries, such as manufacturing skills, marketing and distribution networks and experience with regulatory processes. These distinctive organisational advantages provided the basis for considerable amounts of collaboration between the three groups. Over time, however, larger firms began to develop their R&D skills in biotechnology, and some of the major DBFs began to manufacture and market their own products, so the picture becomes more complicated.

A number of databases show the high level of collaborative activity between firms in biotechnology, and some of the reasons for them. A number of features of collaboration in biotechnology emerge:

– Vertical collaboration between the small new biotechnology firms (DBFs) and larger established firms is common. Pisano, Shan and Teece (1988), taking a random sample of 200 biotechnology collaborative agreements found 62 per cent were between DBFs and established firms; 10 per cent were between established firms; 5 per cent were between DBFs; and the remainder were with universities and research laboratories. This supports the suggestion above of the differing organisational capabilities of the actors involved in biotechnology. A study reported by Yarrow (1988) showed that in 1987 the ten largest US pharmaceutical companies had an average of three partnerships with DBFs.

– The most common area of collaboration is in R&D. Pisano, Shan and Teece found collaboration in R&D in 36 per cent of cases. Hagedoorn and Schakenraad (1990), in a sample of 1213 biotechnology collaborative agreements reported in the press, found joint R&D in 29.8 per cent of their sample.

– The motives for collaboration vary. Hagedoorn and Schakenraad (1990), in a sample of 638 joint biotechnology R&D collaborative arrangements, find the three most common motives to be: technological complementarity (38.1 per cent); reduction of innovation lead times (31 per cent); and lack of financial resources (12.1 per cent). They provide, therefore, a mixture of strategic and tactical concerns. Technological complementarity may be a long-term strategic motive, whereas reducing lead times for particular innovations, or raising finance (by DBFs), may be more immediate concerns. The strategic element of these collaborations may be a reason why, according to Orsenigo (1989), there is a considerable stability in many large/small firm links.

Apart from the increasing expenditure by large firms on biotechnology, many of them have acquired DBFs. 1990 saw Hoffman-La-Roche's majority purchase of Genentech, but earlier Eli Lilley had acquired Hybritech, Bristol-Myers had acquired Genetic Systems, and Glaxo had purchased Biogen's European laboratory.

The Emergence of Biotechnology: Conceptual Background

An idea that can be used to conceptualise the development of biotechnology is that of 'discontinuity'. It will be used here in the sense of a discovery, in this case emanating from the science base, having the potential seriously to affect firms' previous path of technological development. Authors such as Nelson and Winter (1982) and Dosi (1982) refer to the idea of 'technological trajectories'. In the circumstances of specific firms these refer to the way that technology incrementally develops within a firm, and reflects the technology's relationship with endogenous factors, such as the firm's cumulative learning abilities, and exogenous factors, such as market pressures. A discontinuity occurs when a scientific discovery, in this case genetic engineering, profoundly affects existing technological trajectories causing the firm radically to change course.

In the literature on discontinuous innovation (see particularly, Sahal, 1982), these radical changes are assumed to create technologies with marked performance and price advantages over existing technologies. No amount of improvement of the existing technologies can make them competitive with the new product or process. A further element of the literature distinguishes between discontinuities which are 'competence enhancing' and 'competence destroying'. Competence enhancing discontinuities are "order-of-magnitude improvements in price/performance which builds upon existing know-how". Competence destroying discontinuities are "so fundamentally different from previously dominant technologies that the skills and knowledge base required to operate the core technology shift" (Tushman and Anderson, 1987). Hamilton, Vila and Dibner (1990) argue that biotechnology is characterised by competence destruction. However, while this literature is useful in conceptualising the changes that biotechnology has caused, the actual development of the technology highlights the significance of both the continuous and discontinuous elements in innovation.

Biotechnology has the potential to provide price/performance advantages over previous technologies. But, as we shall see below, it has yet to do this. The new exists alongside the old until the new is proven. As Pavitt (1987) says "The commonly held assumption...that new vintages of technology immediately reach economically superior performance is rarely borne out in practice...A period of trial, error, learning and associated incremental change is necessary before a major new technology begins to reach its potential". The new skills of biotechnology may only prove competence destroying when the technology is sufficiently proven and diffused. Indeed, as the new technology is gradually developed, there may well be elements of convergence between it and the older technology. Initially, new activities constitute an additional activity, but if proved successful, they may gradually subsume all other activities, so that a firm's trajectory is radically reshaped.

It is the potentiality of biotechnology to create such a discontinuity – in the sense of in the long run radically altering the way a firm develops and making established competences irrelevant – that provides the background to the industrial interest in it. The scientific discoveries which led to the discontinuity have, in their way, also determined the way that the industrial interest has been shaped. Biotechnology has a number of similarities to Schumpeter's entrepreneurial model of innovation (Freeman, 1982). In this stylised model, science and inventions (which occur outside of industry) attract the attention of entrepreneurs who then induce innovative investment in new technology, and lead to new production patterns and changed market conditions. The attractiveness of the high profits from the entrepreneurial investment causes expansion or 'swarming' of a large number of entrepreneurial firms. These new firms supercede the old, who have to adapt or go out of business. Eventually, with new entrants, profits decline and numbers stabilise.

The 'swarming' effect has occurred, with many new entrepreneurial start-ups being created in the USA and UK during the early 1980s (OTA, 1988; Oakey et al, 1990). The number of new firms started up is, as we have seen, declining, but not because of falling profits, since few firms have achieved profitability, but because of the uncertainties in the technology, and the acquisitive activities of large firms. Indeed, it is the dynamic relationship between large and small firms which provides a closer model for the development of biotechnology. It is akin to Rothwell's (1983) analysis of the growth of semiconductors. The emergence and diffusion of a radically new (discontinuous) technology involves large and small firms engaging in a variety of, sometimes novel, relationships between them.

Biotechnology in the Present Day

A recent OECD report examining the economic and wider impacts of biotechnology (OECD, 1989) considered that although biotechnology has the potential for inducing very radical change, this was still some way from occurring. Using the term 'techno-economic paradigm' (see Perez and Freeman, 1988) the report contrasts the ways in which information technology has profoundly altered the technological system and the economic circumstances in which it operates (causing a change in techno-economic paradigm), whereas biotechnology has yet to have such a radical and pervasive impact. However, the report argues that we are still in the early stages of the full scale application of this revolutionary new technology. It states: "We know from the past experience of the introduction and diffusion of such revolutionary new technologies that...changes in capital stock, in the skill profile and in organisational structures of industry cannot possibly take place in a short period. They are a matter of decades rather than years or months" (OECD, 1989:44).

The report goes on to suggest that biotechnology may become a major basis for new investment and the growth of the economy in the second and third decades of the next century, although in health care this could occur much faster. It considers that by this time biotechnology "will be at the heart of a rapidly growing cluster of new industries and an essential element of competitive survival in an increasing number of established industries" (OECD, 1989:54).

It is certainly fair to say that biotechnology has not to date realised its early expectations of it. In 1988, it was estimated that biopharmaceutical sales amounted to $1 billion of the world industry's $120 billion sales (Financial Times, 10th October, 1989). By the same year only nine human therapeutic products had been approved for marketing in the USA (Bio/technology,

September, 1988); and of the 25,000 active applications for investigation as new drugs, only around 400 are biotechnology-based (OTA, 1988:12). However, four of the 20 new drugs approved in 1986 were biopharmaceuticals (based on rDNA or cell fusion), and this high approval rate is one of the reasons underlying predictions of future rapid growth.

The most complete study of new biotechnology firms in Britain (Oakey et al, 1990) provides a picture of steady rather than rapid growth. In a survey undertaken in 1987/88, only 48 such firms were identified in the UK, and of these, only three employed over 100, and only one (Celltech) employed more than 200. Most of the firms felt that their markets were becoming increasingly competitive, and few were optimistic about future rapid expansion. In the same study six large UK companies, all active investors in biotechnology, were also surveyed. All six companies were investing considerable amounts in biotechnology R&D, and five were increasing their R&D spend in the area. All the large companies expected biotechnology to make an increasing contribution to turnover in the next decade. Three of the large companies were following a deliberate acquisition policy in biotechnology.

The views of some major companies around the world on the potential provided by biotechnology have altered as time has gone by. Sharp (1990) suggests that since 1985/6 most major pharmaceutical companies have been investing to the tune of $100 million or more a year. The Board member responsible for R&D at Hoffman-La Roche has recently stated in the press that his scepticism over biotechnology's ability to yield new pharmaceuticals has been overcome. He now believes that biotechnology has already produced 50-100 potential new drugs worldwide (Financial Times, 20th March, 1990).

Very considerable resources are being dedicated to biotechnology research around the world. It is estimated that in 1988/89 $6 billion was spent on R&D in the US, $3 billion in Europe and $1.9 billion in Japan (Business Week, 26th February, 1990). The number of patent applications derived from this research appears to be increasing rapidly – applications to the US patent office increased from 3,903 in 1984 to 6,907 in 1987, an increase of 77 per cent (Science, 12th February, 1988).

The role that DBFs will play in the future development of biotechnology is very much an open question. Certainly the majority sale of Genentech, by far the most important DBF, has put questions on the ability of DBFs to become fully integrated pharmaceutical companies. The development of drugs, gaining regulatory approval, and their subsequent marketing is a very expensive and time consuming business. It is estimated that the cost of bringing a new drug to market is over $100 million, and a salesforce for a successful drug can number over 100. The financial demands placed on DBFs attempting to market their own drugs may therefore be insurmountable. It is significant that the recent deal between Genentech and Hoffman-La Roche was initiated by Genentech, the largest and wealthiest DBF, seeking new capital to fund new drug developments.

DBFs' creative abilities will in the future, as in the past, be of fundamental importance in deciding their comparative competitiveness, their relationships with other firms, and the role they are to play in the development of biotechnology. It will be DBFs' specific organisational characteristics that enable fast technological learning which will ensure their future. In the Hoffman-La Roche/Genentech arrangement, the primary attraction to the larger company, according to its chief operating officer, was Genentech's scientific excellence and its independence. He believes that it is independence from the unbureaucratic structures and procedures of large companies that is the secret to productivity in drug innovation (Financial Times, 20th March, 1990). Whether this independence can remain after a company has been acquired is an open question. However, the model is there. The future for DBFs lies in their ability to develop dynamic and synergistic relationships with large firms, based on their comparative advantage in technological learning and creativity.

Footnotes

[1] For an excellent exposition of the history of the scientific development of genetic engineering, see Hall (1987).

[2] A 1987 US National Science Foundation survey of 54 companies involved in biotechnology R&D found that over 68 per cent of total R&D was dedicated to health care, and 82 per cent of this was dedicated to therapeutics (Yarrow, 1988). In the UK, a survey of biotechnology firms in 1988 found the largest proportion of them (30 per cent) involved in health care (Walsh, 1988).

[3] The roles played by the Cambridge and Stanford researchers themselves neatly encapsulates perceived wisdom about the differing commercial attitudes of US and British academics. One of the American researchers jointly established a company on the basis of his discovery. The company became the world's most important biotechnology company, and it marketed some of the most important first products to emerge from biotechnology. The British discovery failed, for a variety of reasons, to be patented, and as we shall see it took an initiative from a government agency some years after the discovery to attempt its commercialisation in Britain. The British researchers received Nobel prizes for their work, as did the US researcher who remained in academia.

Chapter 2
Celltech in 1990

Introduction

Celltech is Europe's leading independent DBF. As such it provides an excellent example of a firm seeking to exploit the opportunities provided by biotechnology. Furthermore, it is one of the few DBFs in the world which has adopted a strategy of attempting to become a fully integrated pharmaceutical firm developing, manufacturing and marketing biopharmaceuticals. Like all leading DBFs, Celltech has achieved its position in circumstances of extraordinarily rapid change: in its underlying science and technology base and in the competitive and regulatory strictures under which it operates.

Celltech's story is inseparable from the specifically British circumstances in which it is based. It has to be placed as a new, science-based business in Britain, in the context of: the constraints of financial markets unknowledgeable about the needs of small, high-technology firms; an industry consisting mainly of multinational corporations which initially were generally unreceptive to biotechnology and to new-entrant companies; and an unsupportive public policy environment. On the other hand, it has had the opportunity provided by close links with the scientific excellence (though, in its early stages, commercial inexperience) of Britain's science base.

What follows describes how Celltech has responded and adapted to its changing scientific and competitive situation and analyses these changes by referring to the company's strategy, and particularly to its learning abilities.

Consideration of the various aspects of company strategy, and the factors which have shaped and altered them, provide a major focus of this report. Strategy is seen as an analytical filter linking organisations with the environment – competitive and scientific – in which they operate. Many of Celltech's top decision-makers have held strategic long-term perspectives on the company's development. The following chapters will describe how some elements of the company's strategy, such as its business strategy, have been adaptive in order to respond quickly to environmental change and organisational opportunity. Other aspects of strategy, such as R&D/technological acquisition and human resource strategy, have remained constant so as to assist knowledge of, and influence over, the environment.

Celltech's Current Position

Despite its small size and youth – it employs under 500 people, and is less than ten years old – Celltech has achieved some noticeable successes and maintains extraordinary ambitions. It has undertaken large business contracts with many of the world's most powerful multinational drug companies. The company is developing drugs to treat life-threatening diseases such as cancer, and it has large research programmes in areas of critical care, such as septic shock, and debilitating diseases, such as arthritis and cystic fibrosis. It has been described by the Minister of Trade and Industry as 'the flagship of the UK biotechnology industry'. [1]

The company dominates British independent biotechnology firms. Oakey et al's (1990) survey shows Celltech to be the only such firm employing over 200. The survey shows that by 1988 aggregate employment in UK firms was 1436. With a payroll of nearly 400, Celltech therefore employed rather over one-quarter of the total at this time. The survey also showed an aggregate number of 360 full-time employees in R&D. On these figures, Celltech employed roughly one-third of the total R&D employees in British independent biotechnology firms. [2]

Celltech's Pattern of Growth

Celltech's sales growth in Figure 2, and employment growth in Figure 3, show the typically rapid pattern of growth of a successful start-up company. Data on profits and losses are shown in Figure 4 which shows the company has returned a slight profit since 1987. However, these figures are for profit and losses on ordinary activities after taxation i.e. they include extraordinary items and interest earned. Figure 5 shows that the company has been returning a substantial operating loss since 1987, and the reason for this. This figure, and Figure 6, reveals Celltech's high expenditure on R&D. In 1989, the company spent over £6 million on its own account on R&D, 31 per cent of its total sales. This high expenditure follows the company's decision to adopt a strategy of developing biopharmaceuticals (which will be discussed in later chapters).

R&D expenditure in the world pharmaceutical industry averages at 10 per cent of sales (Walker and Parrish, 1986). A recent US survey of 48 US biotechnology companies (DBFs), found that the average company spent 49 per cent of total operating expenses on R&D (Paine Webber, 1989). So although Celltech's expenditure on R&D is high, it is not comparable with the levels spent in leading US DBFs. This is clearly shown in Table 2 which compares Celltech's R&D expenditure and intensity with some independent US DBFs. Figures are for 1988/89.

Figure 2

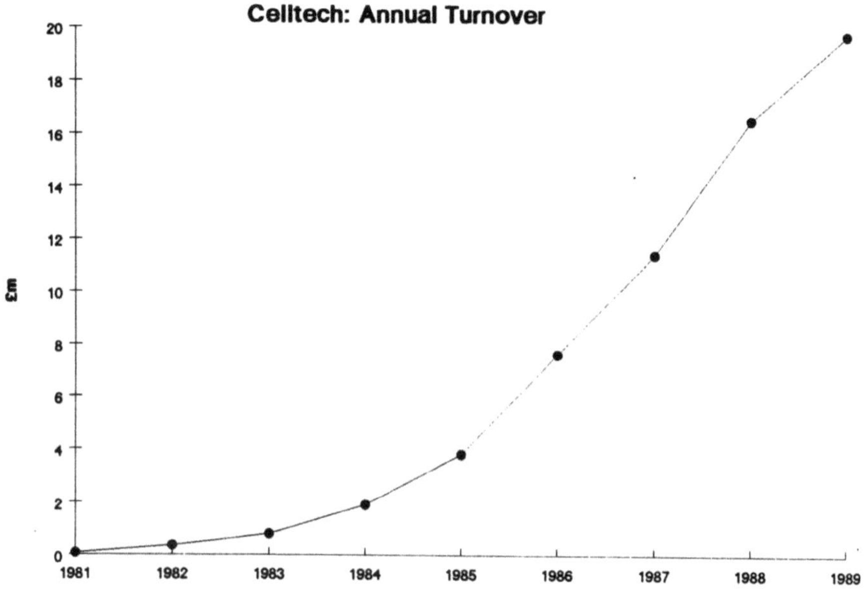

Oakey et al's study shows that only three British DBFs spent more than £500,000 a year on R&D and that Celltech's expenditure is far in excess of that of other independent UK biotechnology firms.

It is not at all unusual for biotechnology firms to be unprofitable for lengthy periods. As a leading venture capitalist lamented recently in the Financial Times (10 October 1989) 'There is not a venture capitalist who has not lost money in biotech'. Paine Webber's (1989) survey of 48 DBFs reported only six firms returning a profit in 1988. The 30 pharmaceutical biotechnology companies surveyed reported between them a loss of $196 million on a turnover of $720 million. Celltech is, therefore, one of the few DBFs in the world to have reported an annual profit.

Celltech's Technology

Biotechnology possesses a number of novel and distinctive features. Although the 'new' biotechnology – focused on the engineering of changes in the genetic structures – incorporates a number of more traditional technologies (Sharp, 1985), much of it is new. This novelty means that many of the techniques used in research and application are experimental. Furthermore, it is a technology which is so qualitatively different from what went before it cannot readily augment previous know-how and practices. Another important feature of biotechnology is its interdisciplinarity. It integrates a wide range of skills and knowledge, including in Celltech's case, for example: molecular biology, microbial and cellular physiology, enzymology, biochemical engineering, separations science, downstream processing and pharmacology.

Celltech has developed an expertise in both rDNA and MAbs elements of biotechnology, and has achieved a level of integration between them. Additionally, the company has become highly skilled at producing and purifying mammalian cell products obtained from these technologies. This technology base provides Celltech with a number of advantages. First, unlike many of its competitors, the company has both rDNA and MAb skills and it is therefore in an advantageous position to integrate the two, for example, in the development of 'antibody engineering'. Some leading US DBFs have concentrated on one of the particular fields of biotechnology. Amgen, for example is primarily a rDNA company, and Centocor primarily a MAb company. Second, it has expertise both in its 'upstream' research efforts and 'downstream' production activities. Third, its production in mammalian cell culture is of superior quality when

Figure 3

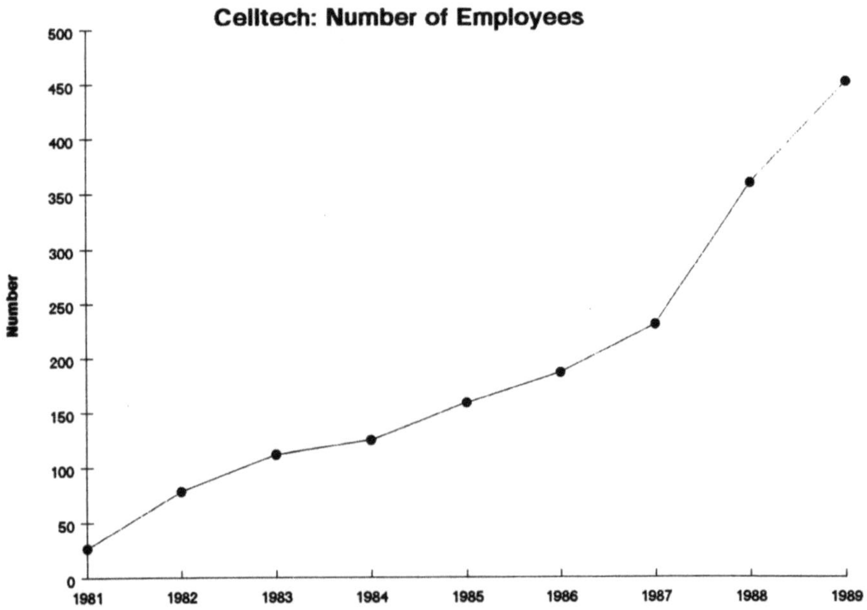

compared to other cell culture systems – such as those based upon bacteria or yeast – used when producing many biopharmaceutical products, particularly large and complex proteins.

Essentially, the company can:

– produce highly valuable, pure proteins from mammalian cells; both MAbs and their derivatives and rDNA modified proteins;

– genetically engineer antibody molecules to make them more effective in human health care.

It is this which provides the basis of Celltech's existing and potential future comparative competitive advantage.

Celltech's Business

Celltech has two distinct businesses based upon its technology. The first is Biopharmaceutical Services. This includes: contract production of MAbs and recombinant proteins; the sponsored development of therapeutics, where the sponsor might fund the R&D and receive marketing rights, or where Celltech undertakes the R&D and sells the marketing rights; and the production and sale of Celltech's own products. The second business is Biopharmaceutical Products. This involves the identification, development, production and eventually the sale

Figure 4

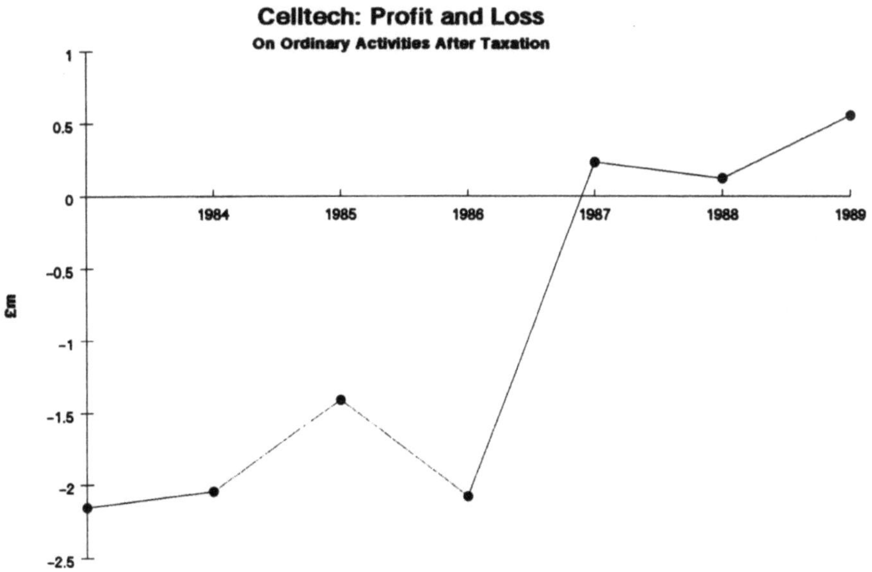

by the company, for its own advantage, of biological molecules for medical treatment. This business is in the process being developed, and at present has no sales.

Biopharmaceutical Services

Contract Production, so-called 'toll manufacture', accounted for around 46 per cent of Celltech's sales in 1988. The company believes itself to be the world's largest manufacturer of high purity MAbs from mammalian cells in culture, and it is also increasingly producing rDNA-derived proteins in bulk. The development of Celltech's manufacturing capacity is discussed in Chapter 7. To date, the company has produced over 6 kilograms of antibody from over 200 cell lines. In biotechnology, bulk production involves miniscule quantities. A few grams product from a batch in a 2000 litre fermenter is considered very satisfactory. Such a batch may take up to three months to process from start to finish.

Figure 5

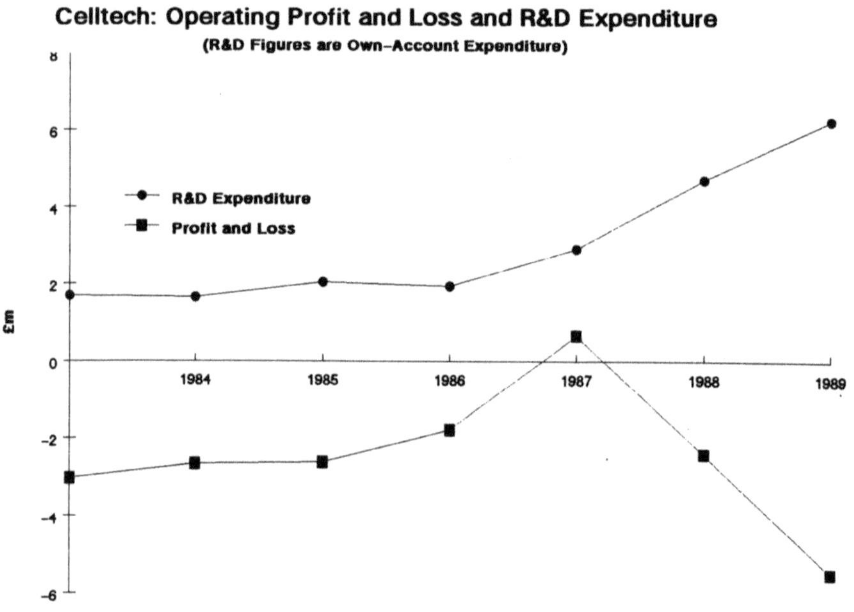

Table 2. R&D Spend and Intensities in Celltech and Four US DBFs

Company	R&D Spend	R&D/Sales	R&D/Employees
Amgen (1989)	£29.2m	66.6%	£57,000
Genentech	£74.5m	50.5%	£43,000
Centocor	£18.6m	54.9%	£43,000
Cetus	£28.7m	70.7%	£31,000
Celltech (1989)	£10.2m	51.9%	£22,000

As an example of this activity, the company provides facilities and services to assist the manufacture of bulk quantities of one of the most important products to emerge in biopharmaceuticals, rDNA-derived Erythropoietin (EPO), for a subsidiary of US Johnson and Johnson (under licence from US Amgen). This dedicated plant began manufacturing in 1989, and added to existing EPO production services for Johnson and Johnson by Celltech (see Chapter 7 for further discussion of the EPO facility). This contract is the largest Celltech has received, and is worth several million pounds.

Contract Development accounted for around 41 per cent of sales in 1988. Typically, the company receives funds for R&D, and payments for meeting

Figure 6

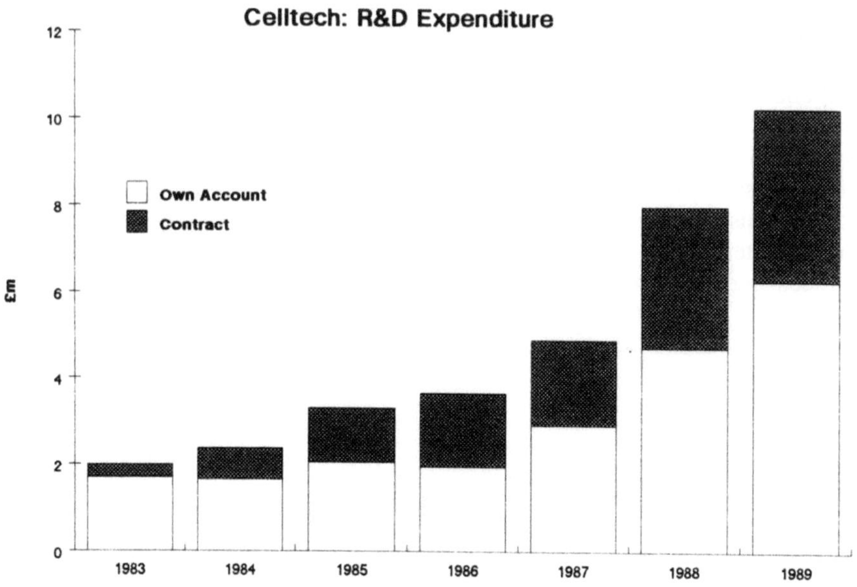

performance targets. It then licenses its discoveries to the sponsor in return for royalties on product sales. Celltech has undertaken contract development for many of the world's largest drug companies, and it has retained manufacturing rights in a number of instances.

Examples in this field of activity include a deal with the UK subsidiary of the American Cyanamid company for the funding of Celltech's MAb-related R&D efforts into the diagnosis and treatment of cancer. This 1986 agreement was renewed in 1988, and has brought over £10 million into the company. In it, Cyanamid receive worldwide marketing rights to Celltech's cancer-related discoveries. Celltech retains rights to all discoveries from the programme unrelated to cancer.

In 1988 a contract was negotiated with the West German Kali Chemie company for the development of human gastric lipase (hGL). This product, which was discovered outside of Celltech, but has been developed by it, has the potential of treating cystic fibrosis sufferers, and if successful will provide the world's first biotechnology-produced drug capable of being delivered orally. (Biopharmaceutical therapeutic agents are proteins. Most normal proteins cannot be taken by mouth because they would not survive exposure to the stomach's enzymes, and would not be absorbed from it. They have, therefore, to be injected. Celltech's work on hGL does not face these problems and provides the potential for an orally-active therapeutic protein.)

In 1989 Celltech and the US company Merck began a collaboration to identify and develop an oral arthritis therapeutic, based on Celltech's enzyme research. By identifying the enzymes that contribute to the process, analysing their activities, and producing inhibitors, the partners hope to be able to retard the progress of the disease, rather than just ameliorate pain and inflammation.

Product Sales accounted for around 12 per cent of Celltech's sales in 1988. These include MAb blood typing reagents, in which Celltech is the major supplier to the US Ortho Diagnostics company, which has around 50 per cent of the world sales in this market. Celltech does not directly sell these products. The company also sells reagents for purifying recombinant and other types of protein such as interferon.

Licensing accounted for around 1 per cent of Celltech's sales in 1988.

Figure 7 shows the trends in the relative importance of product sales, including production contracts, and contract revenues since 1983.

Figure 7

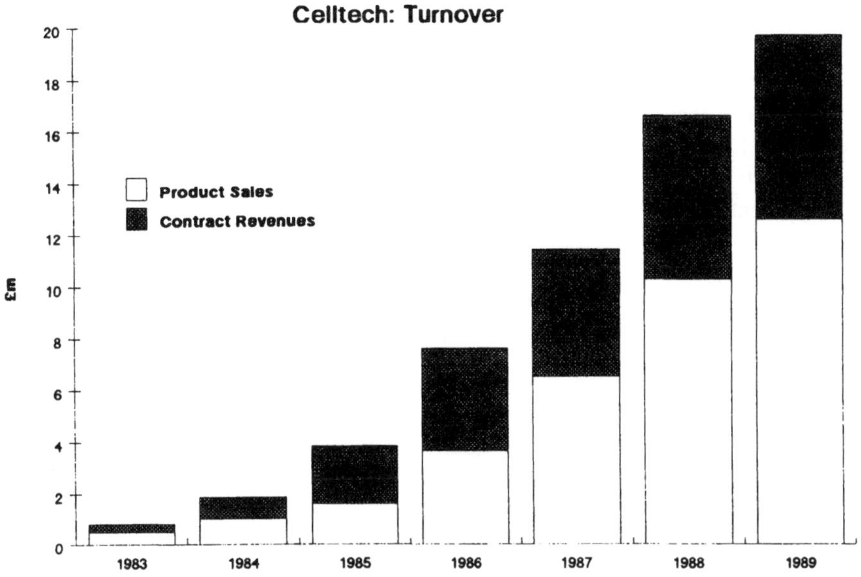

Biopharmaceutical Products

In 1987 Celltech adopted a strategy to produce and sell biopharmaceutical products of its own. The emergence of this strategy is discussed in Chapter 9. Very briefly, it involves developing drugs which Celltech will manufacture and market itself. The drugs being developed are aimed at diseases treated in hospitals rather than by general practitioners, thus requiring a smaller marketing effort. The company is targeting selected European markets for its drug sales. Outside Europe Celltech will license its drugs to pharmaceutical companies when they reach a late stage of development.

The two drugs selected by Celltech as having the most immediate potential are Calcitonin Gene-Related Peptide (CGRP), and anti-Tumour Necrosis Factor (anti-TNF). CGRP is a small peptide which has been found to dilate arteries supplying the brain without lowering blood pressure, and to increase cardiac output with only a small increase in heart rate. In June 1988, the company was awarded its first Clinical Trials Exemption (CTX) for trials in neurosurgery patients suffering a particular form of intracranial haemorrhage. This was extended a few months later to include trials for angina. Anti-TNF is a monoclonal antibody which controls the bacterial infection which can cause septic shock in patients undergoing surgery. It is also believed to have other indications for the control of adverse reactions to some drug therapies for cancer and AIDS patients. Celltech was awarded a CTX for anti-TNF in 1988.

Figure 8 provides data on Celltech's sales broken down by geographical area. It can be seen that the company has a high export orientation. The export sales figures are based on the point at which the goods change ownership, in the case of products, and the commissioning company in the case of sponsored development. Some multinational customers undertake these transactions within the UK, and then themselves export the goods or knowledge within their own organisation. Celltech believes that if the final destination of the business is taken into account the company's export sales are around 90 per cent of total sales. In 1986 Celltech was awarded the Queen's Award for Export Achievement.

Celltech on a World Stage

Celltech is in the opinion of many experts the leading biotechnology firm in Europe. This view is held uniformly by a number of observers of the European biotechnology scene interviewed for this study, including industrialists, policy-

Figure 8

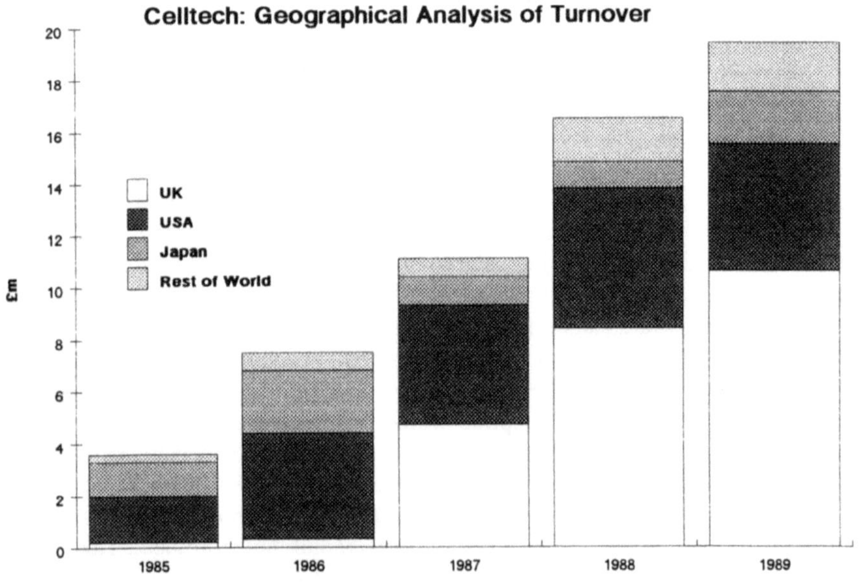

makers and independent consultants. Karl Simpson, the editor of Biotechnology Europe, has recently written 'Celltech is now in a profit situation as (the) world's largest monoclonals producer. It will be the first British start up to make the transition from small high-tech to tomorrow's new style middle size pharmaceutical firm' (Simpson, 1989). We have seen how Celltech dominates British independent biotechnology firms. Two of the highest profile independent UK biotechnology firms apart from Celltech are Porton International and British Bio-technology. The former reported sales of £13.7 million in 1988 and the latter is expected to earn around £3 million in 1989 (Financial Times, 4 November 1989). Perhaps the highest profile Continental independent DBF is Transgene. In 1988 it employed 127 people and reported revenues of FF55 million. It is roughly one-third of the size of Celltech.

While its position in Europe provides the company with considerable status, biotechnology is essentially a global technology operating in global markets. The company has also established a position amongst world-leading biotechnology firms. Celltech's volume of sales places it amongst the top ten DBFs in the world. The Financial Times (10 October 1989) reports the 1988 sales figures of the 'top ten biotechnology companies'. These are shown in Table 3. Had Celltech been considered it would have been included in this list.

Table 3. Top Ten Biotechnology Companies 1988

Company	Total Revenues $m
1. Genentech	334
2. Biosystems	132
3. Cetus	64
4. Centocor	60
5. Amgen	53
6. Diagnostic Products	47
7. Biogen	30
8. Genetics Institute	30
(Celltech)	(29)
9. Collogen	29
10. Genzyme	26

Source: Consulting Resources Corporation.

Table 1 and Table 4 provide some data on the size of a number of high profile independent DBFs. Five of the companies compared are American, including Genentech: until it was recently sold the dominant DBF in the world; Transgene is French, and along with Celltech is amongst the best known DBFs in Europe.

Figure 9 shows the sales growth of Celltech and four reasonably similar sized US firms. These data show that compared to some US competitors Celltech has

Table 4. Total Number of Employees in Some Leading DBFs (1988/89)

Company	Employees
Genentech	1,744
Cetus	770
Centocor	430
Chiron (89)	459
Amgen (89)	537
Celltech (89)	453
Transgene	127

grown less quickly. However, strict comparison of turnover figures is complicated because some US DBFs include R&D expenditure by their Research and Development Limited Partnerships as revenue. On the basis of these data it would appear that while Celltech has not matched the growth of some of the world's leading DBFs, it displays a reasonably similar pattern.

In the final chapter of this report the significance of what Celltech has achieved will be assessed. It will be argued that, despite some very considerable advantages, the company has faced a number of constraints which make its current position something of an achievement. However, proper consideration of Celltech's position amongst world leading biotechnology companies must take

Figure 9

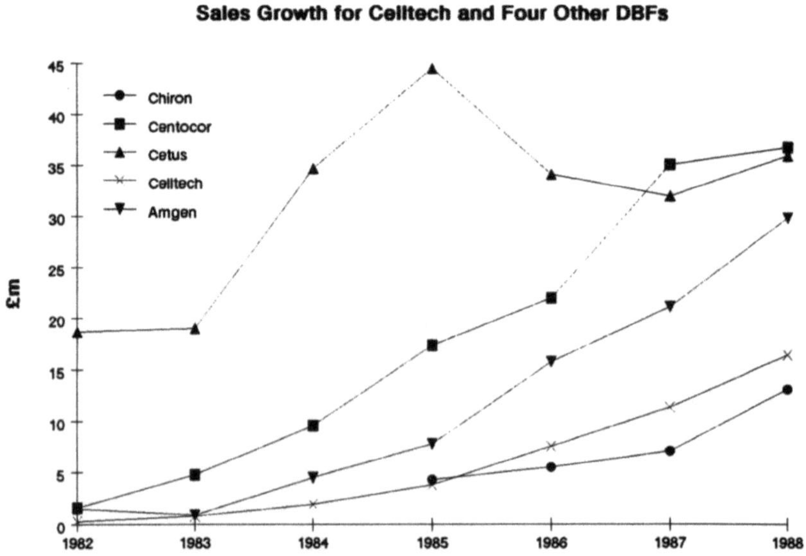

account of its strategy of attempting to develop new drugs. In this respect the company's position can only be considered in the light of the similar strategies of US DBFs, and the strategies of the large pharmaceutical companies which increasingly are using biotechnology in their drug discovery programmes.

Table 2 shows that a number of US competitor DBFs are in a position to spend more on their drug discovery and development programmes compared to Celltech. While this in itself does not show how effective that R&D is likely to be, in the context of biotechnology where competitiveness often depends on the ability to be first to patent, it shows that Celltech may be some way behind the leading US DBFs. Indeed, if we examine the stages of development [3] of a number of independent DBFs' own-product therapeutics, we see that this is the case.

Table 5. DBFs' Development of Own-Product Therapeutics

Company	Pre-clinical Trials	Clinical Trials Phase I-III	Approved
Amgen (1989)	1	8	1
Chiron (1989)	3	7	1
Cetus (1988)	3	5	-
Centocor (1988)	-	3	-
Genentech (1988)	7	4	2
Celltech (1989)	-	2	-

Source: Annual Reports.

Note: These data do not include products developed for other companies. They give no idea of the number of products being developed in R&D.

It should be pointed out that unlike Celltech, which is privately owned, many of the US DBFs are public companies with a broad shareholder base; and all have had access to the sizeable US venture capital market. In the UK venture capital tends not to target technological investments. Whereas 75 per cent of US venture capital investments are believed to be technology related, in the UK this percentage is well below 10 per cent and is declining (Dodgson and Rothwell, 1989a). The situation is similar throughout Europe where in 1988 only £50 million was invested by venture capitalists in the whole of European biotechnology. Only 2 per cent of the total amount invested by European venture capitalists went into biotechnology (European Venture Capital Association, 1989). Celltech has in the past possessed a number of shareholders whose long-term views about their investment has to some extent shielded the company from the short-termism of British financial markets. However, the short-termism of these markets has effectively removed a financing option from the company; an option available to its competitors in the USA.

Comparing Celltech's position with others of the world's leading DBFs Celltech clearly apears to be one of the world's top dozen companies. However, in contrast to some US DBFs, expenditure on R&D and drug development programmes compare unfavourably, and its growth in turnover has been slightly flatter. The reasons for this will be examined in later chapters. What follows is an examination of the factors and circumstances which have led Celltech to its present position, and lead it to believe that it can compete in the pharmaceutical industry.

Footnotes

[1] Lord Young, then Secretary of State for Trade and Industry, Daily Telegraph, 16 April 1988.

[2] The data are for 43 firms, a subset of the total 48 firms identified. Employment in the remaining firms is unlikely to be significant. I am grateful to Sarah Cooper for the provision of these data.

[3] The drug development and regulatory approval process follows a number of stages. Once the substance to be tested has been developed, its potential utility is evaluated in vitro and in animal models. If utility is shown, toxicology studies are undertaken in animals to obtain the approval of medical ethics review bodies and regulatory authorities. These are known as pre-clinical trials, and are followed, provided approval is given, by clinical trials in humans. Stage I of these trials is undertaken in a limited number of healthy volunteers or patients, and the substance is examined for efficacy and safety. If safety and efficacy is demonstrated, Stage II examines the substance in more detail in up to 100 patients. In this Stage further work is done on safety and efficacy, and optimal dosing is considered. Placebo controls are used in this Stage. Phase III, which may involve hundreds or thousands of patients compares the new substance with the best currently approved product.
If these trials show safety and efficacy, a product licence application can be made. All data on the product are submitted to the licensing authorities for approval. The time taken to undertake these trials and get approval varies from seven to ten years, but occasionally can be shortened in the case of drugs for life-threatening diseases.

Chapter 3
Celltech's Genesis

Introduction

The history of Celltech's formation is interesting for a number of reasons. Its creation, and the factors underlying its establishment illustrate well a number of important science, technology and industry policy issues in Britain. However, the pressures which led to its formation were also to influence its subsequent development and pattern of growth. This chapter concentrates on the latter concerns. (For more detailed examination of the former issues of public policy, see Dodgson (1990a and 1990b)).

The negotiations which led to Celltech's formation involved a wide number of government departments and agencies. These included the National Enterprise Board (NEB), National Research Development Corporation (NRDC), Medical Research Council (MRC), Cabinet Office, Department of Industry (DoI), and the Treasury. Also included in negotiations were a number of private-sector investors. The range of interested parties, and the divergence of their views, meant that Celltech's birth was a painful and protracted process. What follows describes and analyses the circumstances which led to the company's formation, the negotiations which established it, and the consequences it had for its subsequent development.

The Background to Celltech's Formation

Developments in Biotechnology

In the late 1970s serious concerns about the state of British biotechnology began to be voiced. Worries were expressed about Britain's inability to commercialise its scientific discoveries in this area. And the US phenomenon of scientists and venture capitalists joining to exploit the potential of the new biotechnology in DBFs seemed to be almost wholly absent in Britain. Most large established companies in both the UK and US were cautious in their approach to this new technology. However, while the number of new biotechnology companies in the US was expanding rapidly: over 50 in existence by 1979, increasing to just under 90 the following year; only a handful had emerged in Britain (OTA, 1988). Faulkner (1988) shows that before 1980 ten new biotechnology firms had been created in the UK, and this number increased to 14 by 1981.

An editorial in Nature in April 1980, headed "Eleventh hour for biotechnology in Britain", summarised the situation. World-leading British science was not being exploited commercially by established firms, neither were new firms being created to do so. It advocated the formation of a state-sponsored biotechnology firm, but it posed the question whether it was too late for this to have any positive effect. It also raised doubts about the role the government might play.

The Conservative Election Victory in 1979

The Tory victory in May 1979 brought to power a government with very strong views on the relationships between government, academia and industry. Mrs Thatcher, unlike her predecessor, assumed the Chair of the Advisory Council for Applied Research and Development (ACARD). According to the Chief Scientist at the Cabinet Office at the time, Professor John Ashworth, the Prime Minister was especially concerned with improving the mechanisms for commercialising British scientific discoveries. The Minister of the Department of Industry, Sir Keith (now Lord) Joseph had similar views, and was well-briefed about developments in British biotechnology. He had visited a number of US biotechnology firms, and had met a number of influential people concerned with its development. High on Joseph's agenda at the time was consideration of the role the NEB was to play under a Conservative administration.

The Role of the National Enterprise Board

Following the 1979 election, the future of the NEB, which had been created by Harold Wilson's government, was uncertain. Its role did not accord with the Conservative ethos of non-interventionism in industry. Indeed, the entire board of the NEB resigned in November 1979 over a political disagreement with Sir Keith Joseph. Throughout its early years in the mid-1970s the NEB had been seen primarily as a supporter of large, and usually ailing, industries. Yet as the case of Inmos, the semiconductor company, reveals, by 1977 it was, in line with its original remit, very receptive to suggestions for supporting small, new technology-based firms. Keith Joseph was very keen to encourage the role the

NEB could play as a catalyst encouraging the development of small, high technology firms, where possible with private sector backing.

By late 1978 a number of officials in the Science Division of the NEB had become convinced of the scientific and commercial potential of biotechnology. These officials, including the Director of the Division, Gerard Fairtlough, began to develop a plan for how UK industry might benefit from this potential. During the early months of 1979, and following discussions with a number of interested parties, a plan had emerged in the NEB to create a small start-up company along the lines of the US model provided by Genentech, Cetus and others. It was designed in the belief that this was the best mechanism rapidly to commercialise scientific discoveries in biotechnology.

One of the first steps the NEB team had taken was to attempt to learn about the scientific advantages the UK possessed in biotechnology. It contracted a US consultant to visit all the UK's major public molecular biology laboratories. He reported the scientific excellence of the world-famous MRC Laboratory of Molecular Biology (LMB) at Cambridge.

The Role of the Medical Research Council

The MRC has played a central role in the world-wide development of biotechnology. The discovery of Monoclonal Antibodies (MAbs) at the LMB at Cambridge is one of the two major breakthoughs underpinning the recent evolution of biotechnology. Senior MRC scientists and administrators claim that during the late 1970s they had unsuccessfully attempted to attract the attention of large established firms to the potentials of its scientific discoveries. The Director of the LMB, Dr Sydney Brenner – for whom the commercial possibilities of biotechnology had been evident for a number of years – was at this time actively seeking a means of commercialising LMB discoveries. When the NEB team told him of their plans he enthusiastically supported them. When the NEB team visited the MRC headquarters in April 1979, they met with a similarly enthusiastic response from its Secretary, Dr (now Sir) James Gowans. The MRC's role in stimulating the creation of Celltech lay in part in its dissatisfaction with the existing mechanism open to it for commercialising discoveries: through the National Research Development Corporation (NRDC).

The National Research Development Corporation

From the 1950s the NRDC had been operating under a system whereby it assumed sole responsibility for commercialising British scientific discoveries from government departments and Research Councils. By 1980 the NRDC was facing criticism from a variety of sources and for a variety of reasons, including being over-bureaucratic, unimaginative, and unable to pick winners. [1]

A particular problem for the Research Councils was that the NRDC was constitutionally unable to pass royalties back to them. So unlike universities which could receive royalties from discoveries patented by the NRDC, and could thus reward individual departments and inventors, research council laboratories, such as the LMB, could not. Royalties from research council-derived patents went into a research council fund, from which scientists applied for funds in the usual way. Within the MRC dissatisfaction with this system was compounded by the belief of many senior officers and scientists that the NRDC operated in a slow and uninventive way. The tensions between the NRDC and MRC were most starkly revealed in a notorious episode of a missed opportunity to patent one of the most influential discoveries in biotechnology: MAbs.

The MAbs Patent Saga

In the middle 1970s, Milstein and Kohler at the LMB developed their Nobel-prize winning method of creating monoclonal antibodies. At the time most British scientists who were receiving government funding were obliged to offer patentable inventions to the NRDC in the first instance. Milstein claims to have informed the appropriate authorities of his research, however, the discovery of MAbs was not patented, causing very critical comment amongst politicians and the press about "glaring mistakes" and the "loss of valuable patent rights". There remains some debate about the timing and extent of information provided to the NRDC, and whether monoclonal antibodies could indeed have been patented. This episode, which resulted in a great deal of soul searching in the NRDC and the MRC, has caused some considerable – and probably pointless – dispute. Much of the importance of this affair lies not with which, if any, organisations were at fault, or what the actual consequences were in lost revenue (if they could in fact be established), but with the way it assumed a symbolic significance as a ready indictment of Britain's inability to commercialise science.

Irrespective of the realities of the situation, the missed opportunity to patent the MAbs discovery provided a high profile example for those, like Mrs Thatcher, concerned with changing the process of technology transfer in Britain. It was with this concern in mind that the Cabinet Office made the case for government investment in Celltech. Professor Ashworth recalls that it was as an experiment in technology transfer that the Prime Minister agreed to the NEB's plans. Lord Joseph also remembers that his primary reason for supporting Celltech was, mindful of the MAbs saga, to change scientists' attitudes towards commercialising discoveries.

The Spinks Report

Very influential on Celltech's formation was the publication in 1980 of a report on biotechnology, which came to be known as the Spinks Report. The Committee which produced the report, which took its name from its Chairman, the former Director of Research at ICI, was established jointly by ACARD, the Royal Society and the Advisory Board for the Research Councils (ABRC). Set up in early 1979, its remit was to study the industrial applications of biotechnology. It reported its findings to the Government in early 1980, and its Report was published that March. Professor John Ashworth, himself an eminent biochemist, put considerable input into the Committee and its Report. He was aware of the NEB's plans, having met with Fairtlough on a number of occasions, and was a staunch supporter of them. The Committee – which included Sir Austin Bide, the Chairman of Glaxo – made 24 recommendations, one of which encouraged the formation of a new biotechnology firm. The Report commented on the absence of new biotechnology firms in the UK, and stated:

"There seems little prospect of similar companies being set up in the United Kingdom at the present time by wholly private finance... We therefore recommend that the National Enterprise Board, in conjunction with NRDC, should investigate the possibility of establishing in the United Kingdom, and with some public funds, a research-oriented biotechnology company of the kind now taking shape elsewhere. A sum of £2 million annually for five years should be sufficient to determine its value and establish the scale of further investment, whether from the private or public sector." (Spinks, 1980:41).

This recommendation provided a focus and greater legitimation for all those concerned with getting Celltech off the ground. It was very helpful to those in the NEB and MRC whose plans increasingly were being entangled in complex and protracted negotiations.

Negotiations on Celltech's Establishment

The original business plan for Celltech had been shaped by the NEB team in early 1979. However, with the resignation of the Board of the NEB, much of the impetus for the creation of the company was lost, and an unavoidable delay ensued. The priorities of the new Chairman, Sir Arthur Knight – as determined by Keith Joseph – lay in dealing with the NEB's involvement with electronics companies. Despite the attraction Celltech provided to Sir Arthur (it was the only significant new initiative the NEB was then considering) a number of months elapsed before the NEB Board considered the biotechnology company proposal. Its approval was gained by March 1980, and the Board formally approved the formation of Celltech in June. The NEB's intention to form a biotechnology company was made public on the 23rd July, 1980.

This lengthy delay placed added pressure on the NEB Celltech team, as they were highly conscious of the lead that US biotechnology companies possessed, and the need to move as quickly as possible. The delay was not entirely wasted time. The NEB continued to develop its plans for Celltech, and the early months of 1980 saw increased negotiating activity. Discussions were undertaken with the MRC about the nature of an agreement between it and the new company, and Draft Heads of Agreement were drawn up in February 1980. The NEB was also pursuing the possibility of reaching some such arrangement with two other laboratories: the Imperial Cancer Research Fund, headed by Sir Walter Bodmer, and Cold Spring Harbor Laboratory on Long Island, headed by James Watson.

Negotiations Between the NEB, MRC and NRDC

Discussions with the NRDC on the NEB's intention to create the biotechnology company began in April. These discussions were to be protracted, and occasionally very heated. To understand why the negotiations between the NEB, MRC and NRDC proved so protracted, one has to comprehend the situation facing these organisations at the time. Rightly or wrongly the NRDC and MRC were facing political pressure over the MAbs episode. As we have seen the NRDC was criticised from a number of sources, and the MRC was facing criticism from organisations such as the ABRC for its perceived inability to commercialise biotechnology discoveries.

The perspectives of the MRC and NRDC on how such commercialisation might take place differed markedly. Recalling previous unsuccessful attempts to excite large firm interest in biotechnology, one MRC Council member remembered that the possibilities were explored of setting up an MRC company to exploit its inventions, before it was discovered that it was prevented by its constitution from doing so. The NRDC, on the other hand, had strong and continuing agreements with the British pharmaceutical industry. One medical discovery, cephalosporin antibiotics, at the time provided the bulk of the NRDC's income. Equally, British pharmaceutical companies were not particularly quick at taking up NRDC licenses, and the majority of its license income at this time came from abroad.

The MRC was therefore very supportive of the NEB's idea of a new, small biotechnology company with priority rights to its discoveries (in MAbs and rDNA) in return for a commitment to commercialise them. To the NRDC, however, it represented a threat to its existing functions and sources of income. The main issue as the NRDC saw it, however, was one of principle, affecting its fundamental position: did the NRDC possess first option rights over MRC discoveries or not, and did the MRC have the right to give them to another (commercial) organisation?

There was a great deal of uncertainty in the NRDC as to whether this development was an isolated case, or one that was likely to set a precedent. After three or four months, however, it became clear that Celltech's formation was inevitable. The NRDC had been handed a fait accompli by the NEB, and the fact that the project had by this time some private sector backing gave it considerable credibility with the government. The NRDC was also still politically vulnerable following the MAbs saga, and it had faced some implied criticism from the Spinks Report, which encouraged it to become more 'entrepreneurial'. Opposing a high profile 'entrepreneurial' initiative designed to prevent future misunderstandings over patenting would not have put the NRDC in a positive light with the government. Having accepted the situation, the NRDC's concern then became how best it might benefit from it.

Discussions between the NEB and NRDC about the type and level of 'compensation' to the NRDC for agreeing to the NEB/MRC initiative took over six months. The Department of Industry became involved in an arbitrator role, but discussions between the two organisations did not proceed smoothly, indeed they were occasionally very heated, and there were some accusations of bad faith in the negotiations. This conflict was inevitable given the positions adopted by the NEB and the NRDC. For the NRDC, the Celltech/MRC Agreement would shut it off from some MRC inventions, adversely affecting future NRDC earning potentials (particularly important given its past reliance on medical discoveries). The NRDC was particularly sensitive about Celltech's plans for a MRC/Warwick University discovery which the company had immediate plans to commercialise. Its view on the situation can be characterised as: "one British company, largely private-owned, not yet formed, with no track record, is to have an inside, exclusive track to public sector research results". For the NEB and Celltech, any 'recompense' for the NRDC would be viewed unfavourably by its private shareholders. Their views on 'compensation' could be characterised as: "involving a loss of income, and it might adversely affect the relationship with the MRC in that it may prove less easy to establish a climate of mutual trust and sensible collaboration". [2]

After months of unsuccessful negotiations the NEB team faced considerable pressures caused by the lengthy delays to the company's formation. A crisis meeting was arranged at the Department of Industry to hammer out a final settlement. After what was reported to be a difficult meeting, an arrangement was established whereby the company was to pay a levy of 1 per cent of turnover to the NRDC, for as long as the MRC agreement was current. This, in addition to a number of royalty agreements related to specific licenses, was considered sufficient by the NRDC to reward it for the loss of its first-right arrangement.

Somewhat paradoxically, given how the NEB's and NRDC's positions were so polarised about Celltech's formation, in July 1981 the two organisations were merged to form the British Technology Group (BTG).

Negotiations Between the MRC and the Treasury

Difficult negotiations had also been taking place between the MRC and the Treasury. Essentially the Treasury was keen to ensure that it controlled any funds that might accrue to the MRC from its arrangement with Celltech. The eventual solution to this set of lengthy negotiations was the establishment of the so-called 'Celltech Fund'. Royalties paid by Celltech on sales and services based on research in MRC establishments were to be paid into the Fund. This Fellowship-awarding Fund was to be administered by a subcommittee of the MRC Council. The Treasury's position was conceded to the extent that the Fund was to receive the first £200,000 from Celltech, 75 per cent of the next tranche up to £1 million and 50 per cent up to £2 million. If the MRC/Celltech arrangement produced bonanza returns (over £2 million), the amount to be paid to the Treasury was to be negotiated.

Negotiations Between the NEB and Private Sector Investors

The government's consent to allow the NEB to invest in Celltech was conditional upon it being formed with majority ownership in private hands. The NEB had, therefore, to attract private financial backers. It unsuccessfully approached a small number of industrial companies for backing. It was then decided to concentrate on raising funds from financial institutions. Through its stockbrokers it produced a shortlist of potential backers, to whom the NEB's business plan for the company, the so-called Blue Book, was sent in June 1980. The Blue Book proposed that authorised capital for Celltech was to be £15m, of which £12 million was to be issued initially, 44% of which to be purchased by the NEB, with four institutions to purchase 14% apiece. Four institutions took up the offer: British and Commonwealth Shipping; Prudential Assurance; Midland Bank and Technical Development Capital, a branch of Investors in Industry. The basic business concept of the company was described as "..a company placed between scientific research organisations and industrial and other 'users': the business will offer developed products and processes derived from original research, and access to leading research teams on a contract research or joint venture basis".

Undoubtedly one of the major attractions of the proposal, and one emphasised in the Blue Book, was the privileged link with the MRC. (The proposed Imperial

Cancer Research Fund and Cold Spring Harbor Laboratory links were included, but nearly all the potential early products identified derived from MRC research). As the Blue Book pointed out, ... "the UK has a relatively under-exploited research base of international quality". And under the proposed Heads of Agreement included in the proposal, investors were offered the opportunity to gain first refusal rights to commercialise the MRC's MAb and rDNA research.

Other factors were important also. Biotechnology had very recently attracted considerable popular attention though the development of interferon, believed to have wide-ranging therapeutic applications. It has also raised interest in the City, with information on the large amounts of venture capital being invested in the US. One of the institutions had already considered investing heavily in a US DBF. There was also, on the part of some of the institutions, an element of investing for national interest reasons.

As the Blue Book proposed, the investors and the NEB established an Organising Committee – chaired by Wynne Denman of British and Commonwealth Shipping – to oversee the formation of the company, including the final approval of the financial structure, organisation and funding. The Committee agreed to fund Celltech with £12 million to be taken up in instalments. This 'drip feed' of capital was designed to ensure that financial control of the company remained with the private sector investors, and that they could direct its business development. At this time concerns were being raised in some quarters about the business potential of a number of NEB investments. These doubts inevitably influenced the approach the institutional investors adopted with the NEB. The NEB itself was in a less than secure position politically. Celltech was the NEB's first major investment where it failed to have a majority shareholding.

Company Leadership

The question of the leadership of the company was one that had been addressed throughout 1980. Three people had been asked to become company chairman: Dr Spinks; the industrialist, Sir Alistair Pilkington; and the renowned Cambridge academic, Sir Michael Stoker, were unsuccessfully approached. Sir Michael did accept a position of some influence in the company which he retained until he retired in 1986.

A headhunter was appointed to find a managing director, and he produced a shortlist of three possible candidates. One of these was initially deemed suitable by the NEB, and he assumed the role of designate managing director. He attended meetings in this role, and for a short time assisted Dr Caroline Vaughan – a NEB official who subsequently worked for Celltech – develop a marketing plan for one particular MRC discovery. However, he did not gain the approval of some of the shareholders and did not take up the post. At this time the NEB chairman, Sir Arthur Knight, made two perspicacious moves. First, he pressed Gerard Fairtlough to assume the role of chief executive of Celltech. Second, he asked John Jackson – a business acquaintance of some years – to become the NEB appointed director to the Board of Celltech. A decade later, Fairtlough was still chief executive, and Jackson was chairman of the company (having taken over this role in 1982).

In addition to Fairtlough and Vaughan, one other person was an employee of Celltech at its formation. Dr Norman Carey, previously Research Director at GD Searle's laboratories at High Wycombe, was appointed R&D director. His appointment, as we shall see in the next chapter, had very significant consequences for the development of the company. In 1990 Carey is Celltech's Director of Scientific Affairs.

A Difficult Birth and an Uncertain Future

When the creation of Celltech was announced in November 1980 it was clear that it had many hard battles ahead of it. It had emerged from a public policy vacuum, as shown by the delays caused by the labyrinthine and protracted negotiations amongst the various government departments involved. In part this was the political consequence of being conceived under a Labour government and born under a Conservative. These circumstances, and the numerous delays caused by the lack of clear policy objectives, led to its emergence as a strange hybrid organisation involving government officials, academics, city institutions and venture capitalists. Whereas the company had a great deal of goodwill, it also faced a considerable amount of scepticism. New Scientist (13 November 1980) commented 'it is difficult not to be sceptical about the staying power and potential of Celltech against competition from the United States and Europe. Its board of directors is dominated by City investors, who, by their own admission, have little experience in assessing the future commercial worth of such esoteric research as genetic engineering'.

Much of this scepticism was justified given the degree of novelty in its aims, attempting as it was to emulate the model provided by the USA. It was endeavouring to transfer technology from a scientific community criticised in the past for its unpreparedness to deal with the business world, to a business environment generally unreceptive to and unsupportive of technological innovation. It was attempting to create a radical new science-based business, using biotechnology, a development which had yet to produce a single marketable product, even in those US and European companies which had a number of years' lead. Moreover, it was attempting to form a new company whose shareholders, as we shall see, held very different views on its aims and its time horizons, particularly concerning their so-called ‘exit routes’. and whose major shareholder, the NEB, was under a certain amount of pressure to disinvest within three years.

The company also had some very positive prospects. It had first-right options to world class science from the MRC, and the active support of the head of the MRC, Sir James Gowans, and the head of the LMB, Dr Sydney Brenner. From the start it was intimately linked with the science base. It also had the world-leading scientific expertise of Dr Norman Carey and some of his erstwhile colleagues who joined him. Importantly, it possessed a basic vision of an exciting opportunity. This vision was created by the NEB and the MRC, and endorsed and supported by a number of exceptionally experienced and successful businessmen, and some eminent scientists, on the Board.

Celltech had emerged from a bruising start. It had challenged a number of long-established practices, and been entangled in protracted and heated negotiations. The extent and convolution of some of these negotiations, and the delays they invoked, would have severely tested a company without exceptional ambition and vision. That Celltech survived this period (though not unscathed), provided testament to the soundness of its belief in the opportunity it possessed, and grounds for some optimism that it might survive the battles ahead.

Footnotes

[1] An article in the New Scientist at the time referred to the NRDC's "time-consuming bureaucracy, lack of imagination, and rigid screening procedures mean many good ideas go begging." (Yanchinski, 1980)

[2] These are actual quotations from internal NRDC and NEB memos.

Chapter 4
The Early Years: Finding A Role; Building a Base

Getting Established

With the battles fought to form Celltech behind them, the company's skeletal staff set about getting established. They moved very quickly. The company had premises within a month, and within a year had attracted a cadre of experienced biotechnology scientists working on Celltech's own R&D efforts. The main thrust of these efforts, and the physical location, were not, however, as envisaged in the original business plan.

Of the three original Celltech employees – Fairtlough, Vaughan and Carey – it was perhaps Dr Carey who was to make the most immediate impact on the company. He engendered a significant change in R&D strategy from the one originally planned, and this affected the location of the company and its early recruitment policy. While the consequences of his appointment were, as we shall see, to cause disquiet amongst some elements within the MRC, his decisions and advice were acted upon, and were profoundly to affect the company's subsequent pattern of growth. In effect, his appointment and subsequent work for the company provided the basis on which Celltech, rather than becoming a research-based technology transfer organisation, adopted a strategy to turn the company into a product-oriented, integrated biotechnology concern.

Celltech's first business plan, the Blue Book, proposed that the company engage primarily in the two basic elements of biotechnology: rDNA and hybridomas. The arrangement with the MRC gave Celltech priority access to its research in both of these areas. Nevertheless, the Blue Book made it clear that the company's first task would be to organise a hybridoma capability. Norman Carey's expertise, however, lay in rDNA. He had made it clear to Board members before his appointment that this was the technology he mainly wished to pursue, and he received their consent. At GD Searle, Norman Carey had developed a world-leading research team in the science of rDNA. The opportunity to recruit into such a small and risky venture an R&D director of Carey's eminence was considered by many on Celltech's Board as something of a coup, not least because he had recently been offered the same post at Genentech.

Before he joined Celltech, a number of GD Searle employees had approached Carey asking to move with him. In view of his plan to increase the emphasis on rDNA work, [1] and the scientific lead his team had established, Carey was keen to recruit some of his ex-colleagues. Within eight months of the company's formation, five scientists had joined him from GD Searle. All five retained senior positions within Celltech in 1990. [2]

The decision to expand the areas of science pursued by the company was not matched by an increase in the level of funding. In effect, the company was expected to pursue two new scientific fields on the basis of the funds deemed suitable for one. This inevitably meant that from the start Celltech was underfunded, and that its management would have very shortly to consider refinancing. This major problem was compounded by the fact that, due to over-optimism on the part of the company, and over-caution towards biotechnology on the part of many large UK firms, the expected research contracts with industry – described in the Blue Book, and planned to produce rapid and significant revenues – were not forthcoming.

The Blue Book stated that the company would probably be located in Cambridge, and a number of articles in the press prior to formation had referred to 'Cambridge-based' Celltech. Fairtlough and Carey examined a number of industrial sites before settling on Slough. The decision to move to Slough was influenced by the fact that it could immediately base the whole company in one building, and that 250 Bath Road included a laboratory which, although imperfect, avoided having to build a new one. The Slough site also had the advantage of being close to High Wycombe where GD Searle's laboratories were based, and where many of its employees intending to move to Celltech lived. By moving to Slough Celltech also displayed its independence from the MRC's LMB. The decision to move to Slough was, however, never intended to be permanent.

Once premises had been secured the company embarked upon two primary objectives: to build up and staff an R&D facility, and quickly to get some products onto the market. Celltech's first Annual Report and Accounts published in December 1981 reported that it had done both: it had furnished modern laboratories, and its first product – antibodies to alpha-interferon – was selling well. The speed with which the former was achieved owed much to the experience of the Searle team, particularly so as the building was an old formulation development laboratory and needed significant changes to turn it into a modern biotechnology laboratory, with its absolute priority placed on clean conditions and on health and safety issues. The latter achievement can be attributed to the fact that it was MRC-derived and had been targeted very early on, before the company had been formed, as a potential product.

Early Products: Success and Failure

In its early years Celltech was very broad-based in its search for commercial possibilities. The Blue Book described the biotechnology market potential in the research market, in health care, industry and agriculture. It did, however, state that 'The initial business strategy of the company will be to develop products based on monoclonal antibodies selling to the healthcare market. Early products will be diagnostics'. At the same time it noted that 'the opportunities available in fine chemicals are thought to be of the same order of magnitude as those in the pharmaceutical industry'. The company explored a number of potential markets outside of healthcare. Indeed, during this period, Caroline Vaughan was quoted as saying that Celltech's main research effort may in the future not be in healthcare but in chemicals (Chemistry and Industry, 6 February 1982), and Norman Carey was reported as saying that Celltech would not become a pharmaceutical company (The Times, 4 August 1981). The breadth of Celltech's search for products has to be seen in the context of the period. Biotechnology was a very new technology, and there was a great deal of uncertainty as to its applications. US biotechnology firms like Genentech and Cetus were believed to have a significant lead in the development of biopharmaceuticals. This led Celltech to explore projects as diverse as metal ore recovery for Shell, and veterinary products for the Agriculture and Food Research Council. However, by the time the 1981 strategic plan was written, four priority projects had emerged; three in the pharmaceutical field, and one in the food industry.

One of these projects – antibodies to alpha-interferon – has already been mentioned. Two Celltech products emerged from the research in this area which had been conducted by Dr Secher at the LMB (who was to become a project manager at Celltech) and Professor Burke at Warwick University. Celltech laid claim to the fact that its product was the only monoclonal antibody to alpha-interferon available commercially in the world. The development of these anti-interferons was very important to Celltech. They meant that from the start the company had products to develop and was not purely a research firm. They also proved to the MRC and its scientists, the Board and to itself that it could rapidly commercialise scientific discoveries.

Another project, involving insulin, was less successful. Despite considerable efforts to attract financial backing for the project from two large pharmaceutical companies, this was not forthcoming and the project folded. The two other projects, blood grouping reagents and chymosin, were to have such an impact on the subsequent development of the company, and reveal so much about the company, that they warrant detailed examination.

Blood grouping reagents (BGRs) [3] were brought to Celltech's attention shortly after its formation by Dr Ed Lennox of the LMB. Dr Lennox, who in 1990 is Director of Research at Celltech, thought the company to be the obvious vehicle for the commercialisation of BGRs. When working on an anti-cancer antibody, Lennox accidently found a MAb which could recognise blood-type A. He offered this discovery to the NRDC which declined to patent it but offered to exploit it by another route. Around the same time he succeeded in the much more difficult task of creating an anti-B reagent. This patent he offered to Celltech, which filed it and was awarded it. Early production of these MAbs was by means of growing them in mice, and was therefore very small-scale. Lennox's team, assisted by others at the LMB, attempted to scale-up production, but this proved very difficult. Eventually a system was established which produced sufficient quantities for their needs at that time. However, when the team subsequently assessed the quality of the system they found it very amateurish. Much greater emphasis, they felt, needed to be placed on developing the production technology and improving instrumentation.

Celltech recognised the potential market for BGRs, which although by its own estimate was not enormous at some £10m a year, was tempting to a company of its size. Dr Birch visited the LMB in the late summer of 1981 and proposed the transfer of Lennox's system to Celltech. This was achieved in practice in three months by two Celltech employees, working seven days a week at the LMB. Once installed at Celltech, the process began of improving this duplicated and relatively simple system. For technical reasons there were advantages in producing blood typing antibodies by cell culture rather than in mouse ascites. Large-scale cell culture had not been attempted before, and so Celltech had to start from scratch. Dr Birch, combining engineering and biology expertise, began work on a system of production using airlift fermenters. Airlift fermenters were chosen rather than other systems because their design was simple, and they had the potential for predictable scale-up. At the same time a systematic study of the physiology of hybridoma cells using this system was undertaken.

By the end of 1982 Celltech had launched two antibodies for ABO blood typing. Within a few years BGRs produced by this method almost replaced those produced by conventional means. While this in itself was an important development for Celltech, as was the production and distribution deal it later negotiated with the world's largest producer of BGRs, Ortho Diagnostics, perhaps the greatest significance of this particular project lay in the fact that it was instrumental in getting the company into the large-scale cell culture business. It had achieved a considerable technical feat, and in doing so had uncovered a demand amongst biotechnology firms for large quantities of antibodies. Celltech benefited commercially from BGRs and also from the credibility it gained in the business and academic community. It also provided the basis, as we shall see, for the company's main short- and medium-term comparative competitive advantage.

Chymosin, the active ingredient of rennin, is the enzyme used to clot milk for cheese-making. In 1980/81 it was thought to have a world market of £60 million. Norman Carey had been aware for some time of a shortage of this enzyme, its relative high value, and the possibility of cloning in this area. Before joining Celltech, he proposed chymosin as one of the projects he wished to undertake. Celltech scientists shared Carey's conviction that it was a logical target to select. It was not a speculative research project inasmuch as it had an identifiable product and market. In early 1981, the project began under the supervision of Dr Mike Doel. The enzyme was cloned, and the project enjoyed some scientific successes. However, the view expressed in the 1981/82 Annual Report and Accounts that 'chymosin produced by recombinant-DNA will reach the market within a few years' was over-optimistic. It was not licensed until late 1986. The project was to provide a number of important lessons for Celltech.

Scientifically, going for chymosin had the advantage of allowing the company to develop its expression systems in E. Coli and yeast. [4] This had continuing benefit for the company in subsequent projects. Celltech obtained two patents from its research, a general and a process-related patent. [5] The first of these was filed on 17 June 1981, and was the company's first derived from its own research. The company was subsequently to learn, however, that a US DBF, Collaborative Research, had a more dominant general patent position. It was found that it was only the process-related patent that had any value. It was following this episode that Celltech learnt the full value of being the first to clone genes, since doing so provides a firm base for a portfolio of patents.

Commercially, the project was less successful in the short term. Over the next few years at least 15 companies were approached to license chymosin, and Celltech came close to licensing it to three. One company was engaged in negotiations for over two years. One of the reasons the licence was not taken up was that there were a number of uncertainties about the final production process. Celltech did not have the funds to invest in developing a final production process, having already spent a significant amount on the research side, and so this element of risk may have deterred potential licensees.

The chymosin saga raised a number of questions within Celltech, most particularly concerning its commercial expertise. The decision to attempt to produce a food industry product, with a team whose experience lay primarily in the pharmaceutical industry, probably has to be seen in the context of the belief in the lead US firms possessed in biopharmaceuticals. However, the company was aware that competitors in chymosin possessed at least a year's lead (Annual Report and Accounts 1981/82, p.11). In an internal document, circulated in July 1985, the company examined the lessons of a number of projects which had not turned out as hoped. It stated that one of the lessons learnt from chymosin was the basic one of the need to fully appraise product market and costs before starting serious project work.

Examining the costs and benefits of the chymosin project is a matter of balancing the high costs, both actual and opportunity (what the resources might otherwise have been used for), against potential revenues and the lessons that the company learnt. On the science side, Celltech cloned its first gene in chymosin;

it achieved its first internally-derived patents; it improved its 'enabling' technology (see Chapter 7) through its use of expression systems; and the project saw the first operational merging of its rDNA and hybridoma competences. Certain lessons were also learnt on the management side, particularly in the need for comprehensive market and economic assessment. Benefits also accrued from developing project management skills and the company learnt some lessons about its patenting strategy. The Blue Book took the view that 'ownership of patents is not regarded as a major competitive factor since there are still legal uncertainties about the patentability of living organisms and in any case there are many potential variants to most biotechnology processes'. Chymosin showed the company the value of patents. In effect, the chymosin project moved Celltech up a learning curve it needed to experience. However, the question to be asked is at what cost?

The total cost of the chymosin project was £390,000. This sum is greater than the company's entire 1982 turnover. The product was eventually licensed in 1986, and so it had market potential, and to some extent was ahead of its time. Despite its excellent science, the market was not ready for it. Furthermore, given the level of expenditure, the product lay in a relatively small market. And, as it turned out, in a market marginal to the company's main efforts. Given the way the company's efforts eventually targeted the pharmaceutical industry, in retrospect it is possible to see the advantages the company might have attained had it selected a health care product. Chymosin was licensed to Pfizer for a lump sum payment and royalties on sales. Early in 1988 an application for recombinant chymosin to be used in cheese-making was filed with the US FDA, the first time a rDNA product was submitted by the food industry.

The chymosin episode gave some ammunition to those on the Board critical of the company's management. For them it highlighted an accusation that was levelled at the company during its early years: that it was commercially naive and its strategy was determined by academic curiosity rather than market need. [6] While the company could argue justifiably that the priority in its early years was to build up its scientific competences, the perceived lack of market focus provided fuel to those shareholders dissatisfied with the way things were going.

Divergent Expectations

From the start there were distinctive schools of thought amongst the shareholders about what the company's objectives were to be, and how these were to be achieved. Different views existed on how, and by when, Celltech was to achieve profitability, and these were manifested in different attitudes towards the company's management. One group of shareholders, typified by the Prudential and British and Commonwealth Shipping, were aware of the long-term nature of investing in biotechnology, and their reasons for investment were directed towards long-term capital accumulation rather than short-term dividends. The other group, typified by Technical Development Capital (TDC), were perhaps more motivated by the fast returns realised by the model of venture capital investment in US electronics companies. An element of such venture capital investment was 'hands-on' intervention in management by the investors. Thus the latter group of investors energetically championed their belief that Celltech needed much greater marketing and commercial input in order to build up a self-sustaining business. TDC representatives were also of the opinion that the company would benefit from building strong trading and shareholding links with major pharmaceutical companies, a view not held by other Board members.

The role the British Technology Group (the NRDC and NEB merged in July 1981) played in this debate is interesting. Paradoxically, given the fact that the NEB had the greatest motive for investing in Celltech for 'national interest' reasons, it was pressing very hard for fast financial returns. While at the same time as it was recording its 'pride' in what Celltech had achieved (see David Fishlock, The Financial Times, 31 March 1982), internal memoranda between senior BTG staff were highly critical of the company, most particularly about its lack of market focus. The BTG was at the time under some pressure to realise some of its investments (see The Economist, 18 August 1984, p.64), and at one time it was considering the possibility of withholding its share of the final tranche of the £12 million. In the event it decided to continue to support Celltech. [7]

The lack of harmony between the company's management and the Board was starkly revealed in early 1982. The company was quickly spending the initial £12 million investment and management had prepared a 'Corporate Planning Document' to attempt to persuade the Board that an increase in total funding of up to £19 million was needed. The company's management argued that the additional investment was required for a new building and a £750,000 fermenter. TDC, in particular, strongly held the view that the company had to demonstrate a tangible success before the £12 million was spent. The Board unanimously decided not to support this request, and in effect told the company to operate within its available resources of £12 million. In November 1982, the Board responded more favourably to a revised 'Corporate Plan', and approved a proposal to seek additional capital. The 1982 Corporate Plan showed some effort to limit expenditure until the new capital had been secured.

At this time Celltech's financial projections began optimistically to suggest high future rates of profit. In 1982 a profit rate of £6.7 million was projected for 1987. This fell to £2.2 million in the 1983 plan, and £1.9 million in the 1984 (in the event, the company made a profit of £0.6 million in 1987). Considering the uncertainty in the business, and with the exception of 1982, Celltech's projections for sales figures have not been too inaccurate. The 1980 to 1986 strategic plans (excluding 1982) have produced figures which on average have been accurate to ± 12 per cent. Figures for profits have been less accurate. However, the figures produced in 1982 were exceptionally inaccurate (for sales as well as profitability). This could be argued to reflect the pressures the managers were under to please the aspirations of some of the shareholders. Having responded to these, the pressures to achieve these unrealistic targets became self-enforcing. However, the decision to attempt to achieve profitability by 1987 was supported throughout the company. It was believed important for the external credibility of the company, and also to be advantageous for its internal management by focusing efforts and building morale.

The picture of Celltech's achievements by middle and late 1982/early 1983 was therefore somewhat mixed. It had built a first-rate scientific capability extremely quickly, and had developed a certain amount of commercial credibility, for which it had received deserved plaudits. Laing and Cruickshank (1982) described the company as 'probably the jewel in the British biotechnology crown'. External perceptions of the company were generally very favourable, both towards its scientific and commercial achievements. For example, Nature (20 August 1981) referred to the company's 'fast start', and to how busy the company was. David Fishlock in The Financial Times (13 May 1983) reported that Celltech 'has harnessed an entire national research effort to its research programme'. And Financial Weekly (12 March 1982) reported that 'Gerard Fairtlough's original conception of the company is already proving sound. A number of approaches have already been received from the organisation wishing to buy equity in the company'. Celltech received a great deal of press attention in its early years and this was assiduously cultivated by Fairtlough and Vaughan.

By November 1982, two years after the company's formation, Celltech employed 110 people, one-third of which were PhD scientists. Despite some less than fruitful projects, such as one involving genome mapping, the company's R&D staff were creatively engaged in some first-rate research, which was to provide a basis for the company's subsequent development. It had successfully transferred technology from academia by commercialising some MRC products, one of which was instrumental in involving Celltech in the business of bulk production of monoclonal antibodies. The company was, however, beset by disunity on the Board and a certain amount of commercial uncertainty. Concerns over its commercial objectives were most strongly expressed by a number of shareholders, but some people working within the company were also questioning the direction the company was taking. [8]

As a result of these concerns, in the middle of 1982 a subcommittee of Wynne Denman, Michael Stoker and Gerard Fairtlough was established by the Board to make clear recommendations on the company's future commercial focus. The committee determined that the company should continue its chymosin work, adopt a project concerned with Acid Stable Lipase, and invest greater resources in hybridomas. The committee also performed an important symbolic role by, for the first time, involving all the constituents of Celltech's Board – shareholders, scientists and managers – in a major decision affecting the company's future.

Creater Coherence and Commercial Credibility

In addition to the results of the formation of the committee, in late 1982/early 1983, a number of events provided greater cohesion and stability to the company. These included changes at Board level, including the chairmanship, and the appointment to the Management Team of a number of people with greater marketing expertise. The latter of these changes greatly assisted the commercial focus of the company. This was achieved through the negotiation of a big contract with a Japanese firm, the formation of a joint venture in diagnostics with a large UK company, and the establishment of a business in bulk production of MAbs.

It will be recalled that before its formation three people were unsuccessfully approached to take on the chairmanship of the company. Eventually, as time was pressing, a representative of one of the shareholders, Wynne Denman, assumed the role on a temporary basis (he agreed with Sir Arthur Knight of the NEB that his first task as chairman was to find his successor). As Denman was an Executive Director of British and Commonwealth Shipping, his time was very limited, and he was conscious of the need for a more active chairman. The chairman's job at Celltech was complicated by the diversity of the Board members and the broadly different organisations they represented. Upon Denman's retirement from the chairmanship on 25 November 1982, Mr John Jackson took over this role. [9] Jackson had been on the Board of the company from the start as the BTG's representative. [10] In his view there was an urgent need for the Board of directors to begin to act in a more coherent and cogent manner, and to build a better working relationship between it and the company's management. Within a few months of his assuming the chairmanship, the Board's structure was changed. The way that Celltech's shareholdership and the financing of the company have altered over time is described in Appendix 3. Two months into his chairmanship, Jackson was informed by TDC of its intention to sell the majority of its equity stake (14 per cent of total company equity) to a large US pharmaceutical company. The view of the Board was that such a sale was inadvisable, as the purchase would allow the US firm to have a seat on the Board, and that it might adversely affect the relationship with the MRC. The company's Articles of Association gave existing shareholders pre-emption rights over sales of shares, but in the event these options were not exercised. The Prudential and British and Commonwealth showed their confidence in Celltech by investing in a new rights issue similar in size to the cost of TDC's shares, and TDC sold its shares to a substantial new investor in Biotechnology Investments Limited (BIL). TDC's sale of shares coincided with the sale of a tranche of BTG's shares (reducing its stake from 44 per cent to 28 per cent). BTG ensured, however, to give the company a public vote of confidence at this time when the company might have appeared vulnerable.

At the same time these changes were taking place (TDC left and BIL joined the Board in June 1983), the organisation of the Board was altered. At its inception, Celltech had a system of Special Directorships which left effective control of the Board with the institutional directors. The City investors introduced this system primarily to ensure that they, rather than the NEB or the company's management, retained control of the company. Decisions by the Special Directors had to be unanimous. They controlled the drip-feed of the £12 million into the company. In practice this allowed the Special Directors to meet before Board meetings to make the important decisions, which then could be presented to the full Board for rubber-stamping, and effectively disenfranchised important constituents of the company such as the scientists on the Board, not to mention the management. The Special Directorships were abolished in June 1983. Also in 1983 Dr Carey joined the Board, extending the representation of management. Within a year, therefore, the new chairman had overseen major changes in the structure of the Board, establishing its basis as a more effective decision-making body. At this time the last tranche of the original £12 million was called up. Also at this time the venue for Board meetings changed from the offices of the Special Directors in rotation, to a neutral venue in a hotel.

Changes were also taking place in the company's management. Until 1982, Celltech's Management Team consisted of Fairtlough, Vaughan and Carey, and the Finance Director, Frits Janssen. During 1982 two more people joined the team (not including Janssen's replacement, Hugh Perrott). David Gration joined from Wyeth Laboratories, where he was Marketing Director (having previously been Head of Pharmaceutical Marketing UK at Boots Company plc), and John Berriman joined from Richard Klinger Limited, where he was Manufacturing Manager (having previously been Sales and Marketing Manager). In 1983 Nic Holladay was recruited from Syntex as Celltech's Director of Health Care Licensing. Holladay went on to become Director of Health Care Marketing in 1984, and Director of Marketing in 1986; he left the company in 1987. As a result of these three appointments, the sales and marketing emphasis on the Management Team changed significantly. Gration also became a member of the Board of Directors.

It was not too long before these changes brought some results. Following a visit to Japan by Gration and Carey, Sankyo, one of Japan's largest pharmaceutical companies, signed two development agreements with Celltech in September 1983. The agreements gave Sankyo exclusive marketing rights to Celltech's research and development work in rDNA-produced tissue plasminogen activator (t-PA) and calcitonin. [11] In return Sankyo paid the full cost of the development of a rDNA production process. Celltech was also to receive success payments for the completion of the projects and receive royalties on any resulting sales. It retained no marketing rights itself. The company estimated the annual market for calcitonin to be £60 million, and for t-PA £330 million.

Those working in the company at the time recall how important this deal with Sankyo was. It raised confidence by proving that Celltech had the credibility to deal scientifically and commercially with an international pharmaceutical company. It also brought in a significant amount of income (the company's contract income increased from £137,000 in 1982 to £396,000 in 1983). The Sankyo deal was also Celltech's first major sale of a R&D package, and provided a model for future deals.

As we have seen, Celltech initially intended first to target products for the diagnostics market. In November 1981, the Management Team proposed the formation of a new Diagnostics venture based on MRC technology. The Board responded somewhat guardedly to this suggestion, and as a result the company decided to recruit a general manager for Diagnostics, and also began to explore the implications of having a partner to provide part of the additional funding required. In November 1982, the Board approved the establishment of a Diagnostics Division, to be headed by David Gration. The Division was to carry out its own R&D, marketing and production. Throughout 1982 negotiations took place with Ciba-Geigy on the possibility of creating a partnership in Diagnostics. These discussions failed, as did talks with Amersham International, and when the Division was formed it was entirely funded by Celltech.

In early 1983 negotiations began with Boots Company plc, and on 12 August 1983 the announcement was made of the formation of Boots-Celltech Diagnostics Limited (B-CD). It was to be a separate, and integrated biodiagnostics company. David Gration became its chief executive officer. Its chairman was Dr Eric Cliffe, a director of the Boots Company, and its deputy chairman was Gerard Fairtlough. Dr Bill Hunter, previously head of the MRC's immuno-assay team at Edinburgh University, became head of R&D. The company was sited alongside Celltech in Bath Road, Slough.

There were a number of reasons behind Celltech's decision to form a joint venture in Diagnostics. Diagnostics were believed to be a rapidly growing market. Celltech's Annual Report and Accounts 1983 reported a prediction of a £1000 million worldwide market in diagnostics by 1991. The obvious attraction of targeting diagnostic products is, as they are used outside the body, they have no need to undergo the lengthy regulatory procedures that in vivo products do. The company believed itself to possess a stream of potential products without the means to market them. It had access to methods of producing diagnostic kits based on MAbs, and believed that MAb technology would come to dominate the diagnostic kit market. A major reason identified by Celltech for the formation of B-CD was that the company needed marketing and distribution to complement its technology. The deal also helped Celltech financially. It realised an exceptional profit of £669,000 from the sale of its Diagnostics Division (the company made an overall loss of over £3 million in 1983). It was also expected that, dependent upon sales performance, additional income of £500,000 would be forthcoming in future years. By forming a joint venture, Celltech also reduced its risk. Also, and very importantly, the company proved that it had science and technology attractive to major UK firms. Doing business with Boots gave Celltech enhanced credibility within the industry, with its scientific collaborators, and with its shareholders.

Celltech's Board was very pleased by the establishment of B-CD, believing that it would provide valuable experience in collaborating with a major company. But while the management and Board were optimistic about B-CD's future, the 1983 Corporate Plan did provide some counter-balancing notes of caution. It observed that the company was sacrificing 50 per cent of the future profits in one of the only two businesses Celltech had so far created with short- and medium-term potential (although of course its capital requirement has halved as well). And, particularly apposite, considering the way things turned out, it pointed out the difficulties of managing any joint venture, and matching the present and future interests of the two parties.

At the same time as Celltech was shaping a business in the diagnostics part of the company, changes were also afoot in the bulk monoclonals business. Celltech's capacity to produce large quantities of MAbs was a direct consequence of its production of BGRs. It was during this period that the company realised that it possessed valuable competences in this area which could be marketed in other ways. On the technical side, the development of the bulk monoclonals business owes much to the foresight and technological lead provided by John Birch. Much of the commercial drive for the development of this business was provided by John Berriman.

Berriman joined Celltech in February 1982, taking charge of administration and engineering. During late 1982/early 1983 he gradually assumed responsibility for developing the bulk monoclonals business. He firstly took over the commercial activities for one of Celltech's proprietorial monoclonal antibodies, NK2 (based on alpha-interferon), and latterly all production activities, including the BGR product line. A visit to the USA in early 1983 convinced Berriman that Celltech had a lead in the bulk production of MAbs, and this provided a significant business opportunity. The realisation within the company of the comparative advantage it possessed in this area manifested itself in the formation on the Culture Products Division in July 1983. (The remaining aspects of Celltech's activities were given the title Healthcare.) Berriman was appointed as the General Manager of the new Division. Celltech's 1983 Annual Report and Accounts claimed that it possessed the world's largest monoclonal antibody production capacity.

By the end of 1983, three years after the company's formation, Celltech was well established at its Slough base, [12] and had a clear scientific and commercial identity. It possessed a significant R&D capacity in both MAbs and rDNA. It had negotiated a deal with a Japanese pharmaceutical company which was to provide the basis and model for other R&D projects in health care. [13] It had formed a joint venture with a large UK firm in diagnostics, and had created a business in the mass production of MAbs. The outcome of these commercial developments were, as we shall see, mixed. The negotiation of deals similar to the ones with Sankyo continued with some success. The business undertaken by the Culture Products Division proved to be highly successful. Sales grew in the Division by 70 per cent compound until it was reorganised in 1986. B-CD was not a success. These outcomes were, however, in the future. What was clear was that by the end of 1983 Celltech as a commercial enterprise had found a role and was rapidly building a base to exploit its increasing scientific strengths. However, despite the changes John Jackson had introduced on the Board of directors – increasing the involvement of its scientist and management members – problems of Board room disunity continued to affect the company.

Footnotes

[1] The company's 1981 Corporate Plan proposed R&D manpower in hybridomas at roughly two-thirds of the level of rDNA R&D manpower.

[2] There was some scepticism about Celltech's ability to attract good scientists. The Listener in May 1981, reported that Celltech 'may have trouble recruiting specialists to service their laboratories' (Dugan, 1981). Maybe it was the credibility that the company gained through recruiting such a highly respected team, but by the end of 1981 the Times Higher Education Supplement (18 December 1981) reported that Celltech was overwhelmed by applications from scientists seeking jobs there.

[3] BGRs are required for the typing of both donors and patients before blood transfusions. Conventionally, typing was done by means of reagents obtained from blood. Celltech's products had the advantage over reagents derived from human sera as they were cheaper, easier to use, more consistent in quality, more accurate, and carried no hepatitis (and subsequently, AIDS) risks to laboratory workers. According to Dr Lennox, their development in the LMB and Celltech and subsequent use owes much to the championing of the product by Douglas Voak of the Cambridge Blood Transfusion Service.

[4] Expression systems are a construction of DNA incorporating the gene coding for a particular substance which has been made so as to maximise production of that substance within the cell.

[5] The first was a general patent relating to rDNA chymosin; the second was a process patent related to the solubilisation of chymosin produced from E. Coli.

[6] Alan Cane, in The Financial Times (16 June 1982), reported (not unfavourably) that 'Celltech as an organisation resembles less a commercial venture than a academic institution'.

[7] In 1983, the BTG set up a company, the Agriculture Genetics Company, which, along the Celltech model, enjoyed exclusive rights to the Agriculture and Food Research Council's research in rDNA.

[8] A number of Celltech employees of the period have told me that they were aware of the need for greater commercial focus at this time. How much this knowledge is compounded by hindsight is, of course, uncertain. The company's Finance Director at the time, Hugh Perrott, remembers 1981-82 being a period of great uncertainty for the company. He recalls great dissatisfaction on the Board, particularly about the lack of clarity in commercial targets. The lack of cohesion in the objectives of the company are perhaps revealed by the fact that before his appointment, Perrot underwent interviews with seven different people.

[9] There was some discussion amongst the shareholders about how Celltech's management might react to a more proactive chairman. It was decided that the benefits a new chairman might bring outweighed any possible adverse reaction from within Celltech.

[10] Jackson was encouraged by Sir Arthur Knight to act as an independent Board member. Indeed, shortly after his appointment to the Board, Sir Arthur resigned from the NEB, and Jackson's direct reporting responsibilities were subsequently never clearly established. The BTG's replacement director was Mr Don Seymour, chairman of LRC International.

[11] Calcitonin is a peptide hormone which regulates the release of calcium from the bone. It is used in the treatment of Paget's disease (abnormal bone formation) and hypercalcaemia (high levels of calcium in the blood), and can possibly be used in treating osteoporosis. The calcitonin gene was isolated as a result of collaborative research between Celltech and Professor MacIntyre of the Royal Postgraduate Medical School and Dr Craig of the Courtauld Institute, Middlesex Hospital (now Head of Biotechnology at ICI). T-PA is a naturally-occurring substance which can only be produced commercially by rDNA techniques. It was expected that Celltech's t-PA would be used for the dissolution of blood clots and for treating athero-sclerosis (hardening of the arteries).

[12] That is not to say that the company had not had any problems about its location. A number of moves had been considered, most particularly to a site on Reading University campus, but these had all been rejected, often after lengthy discussions. Considerations over location were to take up a significant amount of senior management's time in the company's early years.

[13] As to whether Celltech got the best from these deals is uncertain. The company's management was under considerable pressure to attain commercial credibility, and the Board felt it imperative for Celltech to get some deals. This may have reduced the negotiators' ability to bargain from a position of strength. Teitelman (1989) reports how Genetics Systems' first important contract proved considerably more advantageous to its (larger) partner.

Chapter 5
Expanding Ambition

Continuing Commercial and Scientific Development

By 1984 Celltech had identified a number of commercial targets; it had yet, however, to achieve any significant financial returns. Nor did it have any immediate prospect of doing so. Over the next few years the company moved some way towards realising its business objectives. It built upon the base established in the early years by developing the business in Culture Products. It continued with its strategy of joint developments with other companies, on a contract basis or more formally through joint ventures. In addition to Boots-Celltech Diagnostics (B-CD), a venture in microbiology – Apcel Ltd – was established. In 1985 the company began to consider adopting the ambitious strategy of developing its own pharmaceuticals. However, the company continued to face very strong pressure from some of the shareholders emphasising the need to show a return before it could attract further financial support.

The development of the Culture Products Division continued apace throughout the mid 1980s. Details on the company's increased capacity to produce bulk quantities of biotechnology products are included in Chapter 7. The technical successes the company enjoyed are revealed by the Queen's Award for Technological Achievement in 1985 for its work in cell culture technology. In the same year the company was granted the first licence by the US Food and Drug Administration's (FDA) Office of Biologics for the bulk production of MAbs in cell culture. On the commercial side, in 1985 the company reached an agreement with Ortho Diagnostics Systems Inc. (a subsidiary of Johnson and Johnson) to provide them with multi-kilogram quantities of MAbs for use in BGRs to be marketed by Ortho worldwide. This continues to the present day. In 1986 the company received a $1 million contract from Hybritech – a US DBF – to produce a kilogram of MAb.

In its healthcare operations, the company continued identifying new drug opportunities, and in many cases developing them to a stage where a licensee would be interested. The chronology of some of Celltech's major product-based projects is shown in Figure 10. The Figure shows approximate durations of product-based projects in Research and/or Development. Dotted lines indicate periods when activity was not significant.

Following Celltech's successful negotiation of a contract with Sankyo on tPA and calcitonin in 1983, two other smaller, but still multi-million pound, development contracts were signed in 1984 with the same company to produce Tumour Necrosis Factor (TNF) and Macrophage Activating Factor (MAF). [1] These contracts gave Sankyo global marketing rights in return for fully funding Celltech's research programmes in the two areas. The deal included certain 'success payments' for reaching defined research targets, and Sankyo was also to pay royalties on external sales. Celltech was to develop the products up to the provision of manufacturing processes, and provide early clinical trials materials. Sankyo was to be responsible for toxicological work, clinical trials and submissions to world drug regulatory authorities. It was believed that the products would take at least four years to develop, and it was hoped that they would reach the market by 1990. The company believed that if the products were to pass clinical trials then they had huge market potential. Anti-malarial drugs alone were estimated to be worth $200m annually, and the product's total value was put at $1bn per year.

Celltech's four projects for Sankyo were to meet with mixed success. Human calcitonin was successfully produced in bacteria using rDNA techniques, and the technology used was transferred to Sankyo. The tPA project was cancelled by Sankyo in 1987 when the extent of the patenting problems surrounding this product became clear, and the TNF project was dropped very shortly after it began when Genentech patented it. The MAF project was stopped in 1987 when

Figure 10

Product–Based Projects

(Partial List – Excluding 'Contract' Projects)

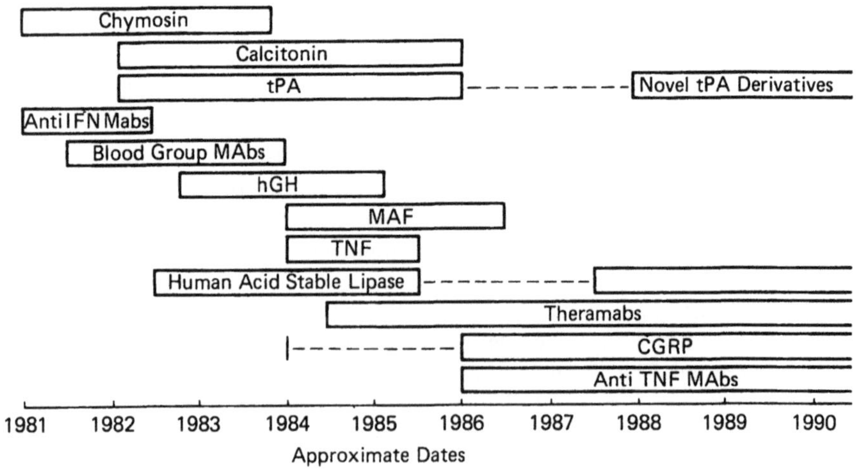

the protein was identified as one already patented elsewhere for other indications.

Also in 1984, Celltech negotiated an agreement with Serono Laboratories Inc. of the USA (a subsidiary of the Swiss-based Ares-Serono group) to produce human growth hormone (hGH). This product was developed to treat dwarfism, and the contract was worth over £1 million to Celltech. Unlike its other deals, in this case Celltech retained the right to market hGH in the UK itself. Additionally, it retained partial manufacturing rights. This product, produced by rDNA techniques, mimicked the natural hormone in human beings thus overcoming the problems then being faced in using actual natural hormone. It also possessed the advantage over competing rDNA products of being produced in cell culture rather than in micro-organisms. This project was successfully completed, and Serono has begun marketing the product. Celltech retained the manufacturing rights of hGH for five years.

Although the company was returning considerable losses, B-CD enjoyed some successes in its early years, launching a number of products, and negotiating two major contracts. Its first product, an alpha feta protein kit used to detect possible abnormality of the foetus, without the risk and expense of amniocentesis, was introduced in 1983. During 1984 it introduced a thyroid stimulating hormone test (reported to do the work of three previous tests and to have a £50 million market worldwide). In 1985 it introduced a diagnostic test for chlamydia (a sexually transmitted disease affecting one million women a year in the UK alone). By the end of 1985, B-CD was able to report in its Annual Report that it had ten products on the market. Also in 1985 it established Boots-Celltech Diagnostics Inc to distribute products in USA.

Two major contracts were negotiated by B-CD during this period. One of these was with LRC International to develop a human female fertility test. [2] LRC funded an R&D programme, the cost of which was estimated to be £650,000, and was to contribute a further £100,000 for start-up costs. The test was to allow women to predict the time of their maximum fertility using a urine sample. It was to be sold over-the-counter, with LRC to have marketing rights over the products in the UK and Europe, and with B-CD having rights in Japan and China. Both companies were to have rights to marketing in the USA and elsewhere. The search for this product was unsuccessful.

The other contract was with Bayer AG of West Germany to develop a MAb-based test to detect the ovulation times of dairy cows. Bayer was to fund the R&D work, in return for world marketing rights. B-CD was to receive royalties on sales of the product which were to be manufactured by Bayer from reagents it supplied. The development of this test was scientifically successful but for a variety of reasons did not reach the market.

Celltech established its second joint venture with Air Products, a US industrial gas and cryogenic plant company, in 1984. The company created, Apcel Ltd., was to undertake R&D contracts in the field of industrial microbiology. Much was expected from the development of industrial microbiology. Celltech's Annual Report and Accounts 1984 stated: 'Industrial microbiology is expected to be the biggest growth area for the application of novel biotechnological techniques by the end of this century, with a very large market potential. The generation of new and improved strains of industrially useful microorganisms is likely to have major implications across a broad spectrum of industries including food manufacture, agriculture, brewing, chemicals, metal extraction and oil pollution control' (Chairman's Statement, p.3). Celltech sold its Industrial Microbiology business to Apcel for £1.7 million. Notably, however, Celltech's rights derived from its chymosin patents were not transferred to the joint company. Apcel was guaranteed contracts of at least £250,000 for the first two years. Its chief executive was Brian Street, Chairman of Air Products (UK), and its chairman was Gerard Fairtlough. The aim of the venture was to develop industrial processes using living organisms applicable to the chemicals, food, bio-polymers, oil recovery, mineral extraction and pollution control sectors. Like B-CD, Apcel was to retain a strong relationship with Celltech, and to forge strong links with academic research.

Celltech's reason for forming the joint venture – apart from the obvious inducement of much needed finance – was that industrial microbiology, in which it had some considerable expertise, took up a lot of management time and resources, and detracted from the development of the Culture Products and Healthcare business. It also noted the potential synergy between Celltech's rDNA skills and the chemical engineering skills of Air Products. Projections for the company were ambitious, with Brian Street claiming that profitability would be achieved within 18 months of the company's birth, and own-products would be launched by year four.[3] Within a year, Apcel had grown to employ 30 people.

A significant change on the scientific side of the company occurred in 1984. A major criticism of Celltech by some of the MRC scientists involved in the company's formation was the lack of priority given to scientific effort in immunology. Indeed a number of people within the company had noted the imbalance in the quality of the research team in immunology compared with rDNA. Whereas the rDNA team was manned by several world-renowned scientists, the immunology team was seen as youthful and inexperienced. This shortcoming was also emphasised on a number of occasions by members of the company's Science Council. Celltech's management claimed, probably with some justification, that it had proved very difficult to attract a senior scientist with these highly-demanded skills. These criticisms were allayed, however, with the appointment of Dr Ed Lennox as Director of Immunology in 1984. [4] Dr Lennox's involvement with Celltech began with the work on BGRs, and during 1983 he worked for the company on a consultancy basis. His decision to leave the MRC (he was approaching retirement age) for Celltech was instrumental in allowing the company to develop its immunology capacities, and in attracting a highly significant contract with American Cyanamid.

The collaboration with American Cyanamid began in March 1986. It was designed to engineer MAbs for the detection and treatment of cancer. Celltech provided Cyanamid – a leading specialist in cancer research – with a programme to re-design antibody molecules. Essentially, Lennox formed and sold a 'package', unavailable elsewhere, where various skills in Celltech were pulled together: purification, expression, large- and small-scale production, and complemented by external expertise. The initial deal was for two years, and was worth £5 million to Celltech. The project is currently in its third stage and has been one of the most important contracts the company has negotiated, both financially and scientifically, particularly in its development of antibody engineering capacities. Celltech has retained the rights to non-cancer discoveries and developments within the programme which has been highly beneficial to its drug development programme.

Continuing Successes, Uncertainty and Divergent Expectations

In its scientific and commercial activities Celltech had achieved some notable successes by 1985/86. It had achieved its ambition of becoming the UK's principal independent biotechnology company, and therefore had a distinct advantage in the sector. To quote Daly (1985:93), '(Celltech) ... is an outstanding example of a European start-up rapidly achieving the technical and commercial success which is associated with American new biotechnology firms'. Fortune International described the company as a 'star', and a 'world leader' in bulk production of MAbs (13 October 1986, p.24). It had a distinctive technological lead in the mass production of MAbs through cell culture, and had important commercial contracts with leading US, Japanese and European drug companies. In its R&D efforts it now possessed front-running expertise in both rDNA and immunology. The company's commercial credibility had been enhanced with the formation of the two joint ventures, which were believed to have high sales potential.

It also possessed some notable weaknesses. Although the company's sales were increasing rapidly, it was yet to realise its cardinal objective stated in all its strategic plans of returning a profit. [5] While the company was confidently predicting profitability by 1987, the situation was highly uncertain. Two potentially high-return business sectors had been formed into joint ventures, and so future returns in these areas were to be halved. In order to fund its discovery work, and to attract immediate income, Celltech had negotiated licensing or royalty deals and retained limited rights itself. It was unlikely, therefore, that the company could achieve the maximum possible return from its investments in R&D.

An additional uncertainty lay in the fact that the scale of any future returns was difficult to assess as the nature of competition – the size of the market and the number of competitors – was unclear due to the rapidly changing technology, patenting and regulatory positions. Celltech's management was aware of its disadvantageous position compared with its primary US competitors. It knew, for example, that Genentech was a formidable competitor with strong patent positions in a range of products and processes in which Celltech was interested, such as tPA, hGH and TNF. It was also aware that the company was spending significantly less on R&D in rDNA than Genentech and Biogen, and was weaker than Hybritech in immunology R&D. Importantly, it was also conscious of how long a lead these companies had, having been established several years before Celltech. Genentech was established in 1976, Biogen in 1978, and Hybritech in 1978. Cetus had been established in 1972. By way of comparison, the two largest independent biotechnology companies, Genentech and Cetus, spent $80 million and $32 million on R&D in 1986; Celltech spent $5 million.

Also at the top of management's mind must have been the knowledge that compared with its major competitors it suffered a significant equity shortage. In the USA, funds of nearly $1 billion had been raised by DBFs between 1983 and 1985. From its inception Celltech's ambition outweighed its financial means. [6]

The company also continued to face disagreements on the Board concerning the nature of the investments necessary to build a sustainable business in biotechnology. These disagreements were particularly apparent during the course of discussions about two instances when the company's management requested additional investment. A certain amount of concern was inevitable about such investment. By 1985 Celltech had been operating for five years, and had no prospect of showing any return for another two. As shown in Figure 5, heavy R&D spend was suppressing profitability. The company's major own-risk project, chymosin, had proved a commercial failure. Yet from the start it was accepted by some Board members, including the representatives of the Prudential and British and Commonwealth Shipping, that biotechnology was a long-term investment, and they adopted the strategy of backing the management on the basis of their assessment of its espoused commercial and technological strategy. As the technology was developing so rapidly, part of this strategy occasionally meant that the company had to respond quickly to opportunities, and this required additional investment. However, some Board members were unconvinced about the case put forward by management for supporting such investment.

One example related to a proposed investment in a purpose-built plant in a nearby building for the mass production of MAbs. A lucrative short- and medium-term market had been identified in this area, and it was the management's plan to continue to expand in it, in the belief that it might turn into the company's 'cash cow'. A number of assumptions underlay management's strategy for bulk production of MAbs. It was believed that cell culture would become an important technology, that it was unlikely to be improved upon by competitors, that regulatory approval would be forthcoming, and that the company's head start would give it a distinct advantage over US and Japanese manufacturers. To expand, the company needed to invest in a new building and plant, and this was estimated to cost just under £5 million.

This investment – which was first proposed in 1984, outside the corporate planning procedure – was opposed by a number of Board members representing shareholders. Particularly antipathetic to the investment was BIL's representative, Sir Francis (now Lord) Tombs. [7] Apparently Tombs felt that such expenditure should have been dealt with in the annual corporate planning exercise, and greater emphasis should have been placed on assessing alternatives. The differing views on this matter were eventually reconciled by the chairman following a number of meetings and the provision of further position papers by management. Eventually, unanimous Board approval was given to go ahead with the investment. [8] The investment in the plant proceeded successfully, and the company reached profitability in 1987.

Another disagreement arose some 18 months later, in early 1986, about a plan put forward by management to construct a new building where its main offices and R&D laboratory could be located. Lord Tombs was of the opinion that the burden of the proposed expenditure was likely to have the undesirable effect of pushing back the company's profit horizon. He argued that the timing of the move had been influenced by the availability of the proposed site, and that the company's prospects were insufficiently certain to justify either a very substantial increase in its fixed overheads or the imposition of a major strain on its balance sheet. The Board of directors, however, supported management's plan, and finding himself in a minority of one, Tombs resigned. BIL continued its investment in Celltech, although it did not formally appoint a replacement for Tombs on the Board. It eventually sold its shares in 1987.

BIL's dealings with Celltech remain a curious episode. Biotechnology Investments Limited was established by Lord Rothschild with the explicit purpose of investing in promising biotechnology companies. It was reported to have found some difficulty in finding UK companies which matched its investment criteria. When in 1983 BIL invested £2.8m in Celltech for a 14 per cent stake, it was its first investment in an unquoted UK firm. One report stated that BIL was a reluctant investor, and invested in Celltech not through choice but through establishment pressure (Financial Weekly, 26 February 1987, p.28). Whatever the reason, involvement of the high profile investor with Celltech was beneficial to the company. The Economist (18 June 1983) reported that 'The investment will certainly enhance its credibility when more capital is required'. BIL's chosen Board representative, Francis Tombs was, and is, on the Board of NM Rothschild, the company behind BIL. His appointment to the Board of

Celltech was made despite the bulk of his highly successful career being as a leader of large industrial organisations (as chairman of the engineering companies Rolls-Royce and Turner and Newall, and he was previously head of the Electricity Council). BIL was an investor in Celltech for four years. It made a profit of £6.4 million on its investment of £2.8 million. Shares in the company having risen from £1.75 to £5.17 between 1983 and 1987.

Towards the Pharmaceutical Strategy

It was in these circumstances – with the company showing some strengths but a number of weaknesses – that Celltech began to consider its boldest move to date. A Board Discussion Paper, circulated in early 1985, raised the question of whether Celltech should itself attempt to become an integrated pharmaceutical company. The Discussion Paper described major uncertainties over the company's ability to attract the large investment needed to realise the strategy, and there was considerable doubt over the nature of competition in biopharmaceuticals. Nevertheless, the company's management recommended that the pharmaceutical strategy be adopted. It was probably inevitable, given Gerard Fairtlough's ambition for the company and Norman Carey's pharmaceutical expertise, that such a strategy would be adopted. (It was also David Gration's opinion from the day he joined the company that it should go for drugs.) Whether this was the case or not, it was not until this time that the company had the confidence to make this ambition manifest. The pharmaceutical strategy was formally adopted by the Board on 19 March 1987. This Discussion Paper will be considered in greater depth in chapter 9.

Two important aspects of Celltech's achievements underlay the confidence with which it considered attempting to become a pharmaceutical company. First, the company had highly advanced scientific and technological skills in R&D and manufacturing. These resided within the company, and within its ability to work collaboratively with leading UK academic scientists. Second, through careful cultivation of an organisational culture conducive and attractive to creative and highly motivated employees, it believed it could produce a working environment which could compete with large multinational companies. These are examined in more detail in Chapters 6 and 7.

Company Reorganisation

An important change in the company's management structure was to occur in Spring 1986 with David Gration returning from B-CD to take up the post of Commercial Director. In April 1987 he became Celltech's chief operating officer assuming responsibility for much of the day-to-day management of Celltech. Gration instigated a number of significant changes in the organisation of the company. One of his immediate concerns lay with the Culture Products Division which had become slightly remote from the rest of the company. To overcome this he reorganised the company on a functional basis, similar to the present structure (see Figure 11). The General Manager of the Culture Products Division, John Berriman, became Director of Sales, and subsequently was made Director of Sales and Marketing in Summer 1987. A Director of Manufacturing, Mike Harvey, was appointed in December 1986. Ed Lennox became Director of Research and Academic Liaison, and John Birch, Director of Development. The functional directors, including the Finance Manager, reported to David Gration.

Other important organisational changes took place during this period. In September 1987, John Huckle joined Celltech as its Finance Director, joining Fairtlough, Carey and Gration as an executive Board director. On the Research and Development side, in 1988 Norman Carey was created Director of Science, responsible for ensuring the continuing scientific development of the company, academic linkages and patenting.

Into Profitability and Successful New Deals: Out of Joint Ventures

1987 was an important year for Celltech. It made a profit for the first time (see Figure 5), it negotiated an important manufacturing contract with Johnson & Johnson, and the Board gave formal approval to the pharmaceutical strategy.

Figure 11

Executive Functional Structure

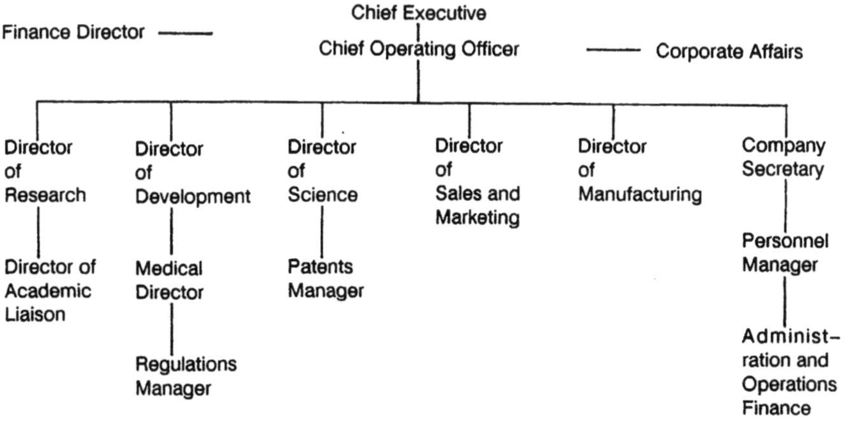

Major developments also took place on the financial side. In August 1987 BIL sold its shares in Celltech. The loss of a major, and high profile investor such as BIL was a considerable set-back for Celltech, and could have adversely affected its ability to raise further equity from the City, particularly as in 1987 the company undertook a private share placing to fund the pharmaceutical strategy. Fortunately for the company in the short term, this did not turn out to be the case, and British and Commonwealth Shipping took the opportunity to increase substantially its shareholding.

The initial document discussing the pharmaceutical strategy estimated that it would require funding of around £60 million. The 1987 placing intended to raise £45 million, although the option existed for this to increase to £75 million if the demand existed. The timing of the placement, less than two months after the Stock Exchange Crash of October 1987, was, to say the least, extremely unfortunate for the company, and the amount raised was disappointing. Biotechnology firms were badly hit by the crash on both sides of the Atlantic. In the two weeks from 12-28 October 1987 the stock value of 60 leading biotechnology companies fell 44 per cent (compared to a fall of 28 per cent for the 400 largest US companies) (Biotechnology Bulletin, Vol.6, No.12, January 1988). Celltech had hoped prior to the crash to sell its shares at £6.00; this fell to £4.00 after the events in October. In the event, Celltech raised £42 million, £24 million from existing shareholders, around £8 million from Japanese companies, and nothing from the USA. [9] Although the company was unable to raise as much as it hoped, the fact that it got near so close to the Crash is testament to the attractiveness of the offer. It was also at the time the largest private placing ever to have occurred in the UK. The finance raising process was an important learning exercise for the company, with its very detailed demands for information requiring a significant amount of management and Board member time.

As the pharmaceutical aspect of Celltech's strategy continued to develop apace, it was becoming apparent that another earlier aspect of strategy was going wrong. Apcel and B-CD were not realising their initial expectations in terms of profits. Apcel went into voluntary liquidation in 1987; in the same year B-CD made a loss of nearly £2.5 million, and Boots Company informed Celltech of its decision to sell its share. Data on B-CD's turnover and losses are provided in Table 6. These show the extent of the problem.

Apcel's business was actually developing reasonably well. Sales doubled between 1985 and 1986, from £0.6 million to £1.3 million; the number of employees increased from 28 to 39, and its losses these years were £250,000 and £275,000 respectively. However, from the start there were differences in the perspectives of the two shareholders. According to managers from Celltech's side, a problem arose over attempts to adopt a flexible commercial strategy. Air Products were more used to a formal 'blueprint' for commercial development, rather than the entrepreneurial, adaptive strategy used by Celltech. Apart from internal disagreements, Air Products' corporate decision-makers were reassess-

Table 6. Boots-Celltech Diagnostics: Turnover and Losses (£000s)

	1984	1985	1986	1987
Turnover	376	1461	1982	2367
Losses	(1573)	(1112)	(1882)	(2447)

ing its diversification strategy, and external consultants had recommended to the company that it divest its biotechnology interests. Celltech acquired the business of Apcel on 1 April 1987 for £156,000.

B-CD was a much larger venture, and as it was hoped that it would play a major role in contributing to Celltech's profitability its eventual demise was therefore much more of a problem. The decision by Boots to dispose of its shares was described in Celltech's 1987 Annual Report and Accounts as a result of a strategic decision to concentrate on its core businesses. There are divergent views as to the cause of B-CD's failure within Celltech. These include: lack of understanding of the marketing requirements of diagnostic products; remoteness from Celltech and its science and technology base; rushing into US markets too quickly; lack of continuity in management; and mismatches in expectations from the partners on what the company was and what it was to do. Given the diversity of opinion on the causes of its failure it is difficult to assess the actual reasons. However, it is clear that the company lacked managerial leadership during crucial periods. There was, for example, delay following David Gration's rejoining Celltech before his successor, could begin to operate effectively. This inevitably disrupted the company's momentum and development. Celltech acquired Boots' shares back in March 1988 for £2.5 million, and B-CD was integrated into Celltech, trading from January 1989 under the name of Celltech Diagnostics Ltd. The major part of this business, diagnostic kits, was sold to Novo in March 1989. Part of B-CD's portfolio, its 'strips' technology, was retained by Celltech and has subsequently been licensed out.

Rapid Changes in Biotechnology Markets

To conclude this chapter it is worth re-emphasising the way that Celltech's commercial development, described in the last two chapters, has necessitated rapid change in commercial objectives. Illustration of this is provided by two examples of the speed and comprehensiveness of changes in the company's commercial opportunities. The first relates to the balance in the importance of MAb and rDNA technologies. As we have seen, to date only nine biotechnology therapeutics have been developed, and of these only one is MAb based. There have been a number of delays in establishing the clinical efficacy of MAbs, and there have also been problems of immune response. This has meant that the market for MAbs-based therapeutics, which was predicted to be a significant proportion of biotechnology product markets, has grown more slowly than was expected. For Celltech this has meant the need to transfer the skills initially developed for MAb production to rDNA protein production. Contract production of rDNA products is currently three times higher than contract production of MAb products, and this differential is likely to increase. Bulk production within Celltech has also altered over time from concentrating on diagnostics to therapeutics. The latter has, of course, far greater demands over standards.

The other example relates to the development of biopharmaceuticals. The 1987 prospectus listed six products which Celltech was in the process of developing. Within two years the situation has changed considerably. Of the six, two have subsequently been discontinued because of scientific obstacles, one has been incorporated into a collaborative arrangement where Celltech retains no marketing rights, and three continue with two in particular showing increasingly favourable indications. Of the three continuing projects, in two cases uncertainties remain over the company's patenting position. The ways that Celltech has attempted to deal with such rapid and disruptive change are the focus of the next four chapters.

Footnotes

[1] TNF was believed to have the capability of destroying tumour cells and malarial parasites, while MAF was believed to stimulate the body's immune system to kill some tumour cells, inhibit their growth, and also have the potential for treating prolonged infections such as tuberculosis.

[2] The test was invented by Dr Terry Baker and Dr Bill Coulson at the Middlesex Hospital Medical School. Dr Baker became B-CD's R&D Director when Dr Hunter retired in 1984.

[3] Interview with Brian Street, Management Today, March 1985.

[4] One other senior management change of note took place in 1984 when one of the original executives, Caroline Vaughan, left to pursue a career as a venture capitalist.

[5] The 1980 Blue Book – the first business plan – estimated that the company would return a profit by 1985. This assumption carried through into the 1981 plan. The 1982 plan suggested a profit by 1986, and the 1983 plan revised this to 1987.

[6] An internal BTG memo, evaluating Celltech's performance in 1982, stated that the company was 'ludicrously under-funded' for its attempt to establish a position in the biotechnology market.

[7] Lord Tombs declined to be interviewed for my study of Celltech, the only person to do so. I report his views as described by six Board members and attendees. Apparently Tombs was strongly of the opinion that Celltech needed to achieve a profit so as to increase external credibility, and also to assist effective internal

management. BIL's alternate on the Board, David Leathers, held similar views to Tombs.

[8] It is unlikely that the decision to make this investment would have been made by the Board under the previous system of Special Directorships.
Following some criticisms by Board members of Celltech management during this episode, Gerard Fairtlough acknowledged at a Board meeting in September 1984 that these had been useful and had been accepted. Typically, he also noted that risk aversion was to be avoided, and to quote the minutes, he stated: 'To achieve success in competition with US and Japanese companies Celltech needed imagination, calculated risk-taking, flexibility and ability to learn, and speed in taking corrective action when necessary'.

[9] Celltech's Finance Director, John Huckle, reported that in the UK, presentations were made to 250 potential new investors, with 35 subsequently investing. In Japan presentations were made to 14 investors, eight of which subscribed (Genetic Engineering News, Vol.8, No.2, February 1988).

Chapter 6
Building a Creative Organisation

Introduction

Celltech was established, and has grown, on the basis of its ability to act creatively and quickly. Underpinning this ability has been a high level of dedication by senior management to the creation of new forms of employment system and organisational structure to ensure that these abilities are retained as the company grows. This chapter will describe and begin to analyse how Celltech built a creative organisation. The first part considers how the company attracts and retains high quality staff, the second part considers the organisational structures in which they work.

The Employment System

Celltech's top management has placed great emphasis on policies designed to recruit and motivate talented company employees. This has been a key feature of strategy right from the company's inception. Its original business plan – the NEB's 1980 'Blue Book' – envisaged that the company's chief executive would take personal responsibility for 'two important areas': planning and corporate strategy, and career development planning. It went on to say 'The major asset of the company will be the quality of its people'. Since then, in its Annual Strategic Plans, the company has repeated that amongst its primary objectives has been 'recruiting first-rate staff and developing them to their full potential'. That these are not platitudes, and are actually acted upon, owes much to the leadership style and beliefs of Gerard Fairtlough, but his views appear to have infused throughout Celltech's management. As the company has grown there has been an increasing 'professionalisation' of the personnel function, and instead of almost entirely recruiting the required skills, greater emphasis has been placed on training people in-house.

Celltech's Employees

The growth in the number of employees in Celltech up to 1990 is shown in Figure 4. The company's employees are characterised by a number of exceptional features, including high educational qualifications and young age profiles. At the beginning of 1989, the company employed 88 PhD scientists, and 131 graduates or postgraduates. Over 55 per cent of the company's staff, therefore, were graduates or had attained postgraduate degrees. At the same time, three-quarters of Celltech's employees were under 35 years of age, and only 12 per cent of employees were over 40. Figure 12 shows the age distribution of Celltech's employees. There is also some limited evidence to show that the company has enjoyed a low level of labour turnover. Figure 3 shows the rapid increase in recruitment in the late 1980s. A high percentage of staff, have, however, been with the company a comparatively long time. Over a quarter of Celltech employees employed at the beginning of 1989, joined the company in 1984 or before. Of the 125 employees in 1984, a very high 101 still remain in the company.

Figure 12

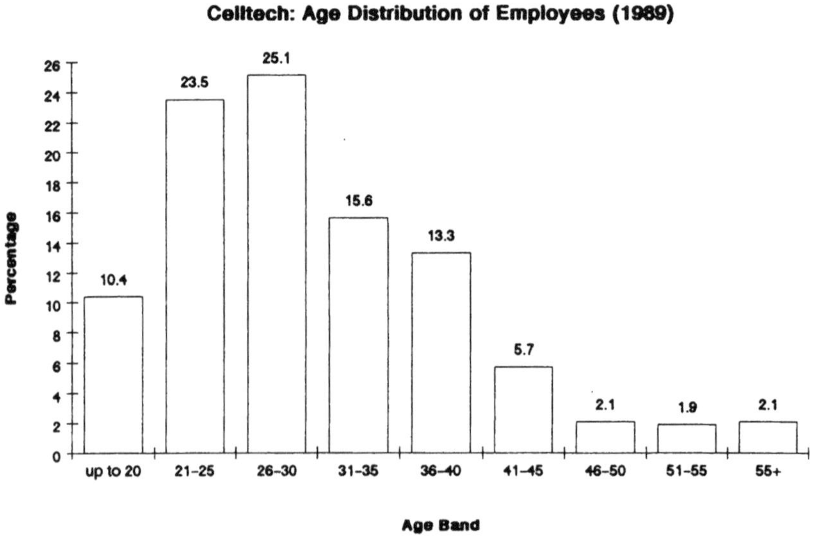

The Personnel Function

Trade unions are not represented in negotiations at Celltech. This gives the company's Personnel function considerable leeway in its management-labour relations. [1] Celltech's Personnel Manager, Peter Nicholls, joined the company in 1987. He has a staff of six, including a Training and Development Manager, and a Recruitment Officer. The following briefly describes some of the functions of the Personnel Department; including administering the payment system, training and recruitment.

Celltech's Payment System

There is no system of job grading at Celltech, staff are graded individually. There are, however, broad classifications of grades in which individuals are placed. [2] Salary grades and ranges are generally published. Despite the hugely time-consuming process of individual assessment, this system is preferred as it enables considerably greater flexibility in allocating rewards. Despite some inevitable mistakes, Peter Nicholls believes that the system has achieved its avowed objective of creating a meritocracy. It is his intention to delay the introduction of a job classification system for as long as possible.

Attempts are made to keep salaries competitive with a number of firms working in similar areas, such as ICI and Beecham. In 1988, 16 per cent was added to Celltech's salary costs. This was added mainly to the lower grades, which were believed to be comparatively underpaid.

The company has considered the use of a profit-related incentive scheme. Indeed, bonus payments were made for achieving profitability in 1987. However, as profits have been low or non-existent, this reward scheme was considered ineffective, and has not been continued. Other incentive schemes, such as reward in line with sales growth, have also been considered and rejected. A management bonus scheme is in place. The amount managers are awarded through their bonus scheme is not declared.

Celltech operates three share option schemes: the Celltech Executive Share Option Scheme, introduced in 1985, for senior staff; the Celltech Savings Related Share Option, for all staff, introduced in 1987, allows any employee to save up to £100 a month, which can then be invested along with interest earned in future options; and the Celltech Employee Share Scheme which allocates shares to employees if certain profitability targets are met. These schemes together have the potential to offer up to 12.5 per cent of total company issued share capital. In 1987 the company's employees, directors and Science Council members held 3.5 per cent of ordinary share capital.

The executive share options are offered in part in order to compete with US companies such as Genentech, Cetus and Biogen which offer generous option packages. It is a mechanism primarily for retaining key scientists and managers. On a number of occasions Celltech has allowed departing managers and scientists to retain their options as a reward for their efforts during their time with the company. An additional incentive scheme involves quarterly bonus awards for those considered to have made exceptional contributions. The awards, which are not open to senior staff, offer one gold award of £1000, two silver awards of £500 each and four bronze awards of £250 each. The recipients can be nominated by anyone and are assessed by a panel representing a cross-section of the company.

Appraisal System

There is a formal appraisal system undertaken for all staff to inform an Annual Review undertaken in July of each year. The appraisal is compiled by each employee's superior with the active involvement of the employee. Every employee must see, and agree with, his or her appraisal. Quarterly objectives are agreed, and past performance in achieving these objectives assessed. Emphasis is placed on rewarding enthusiasm and commitment. This inevitably is a time-consuming process for the appraisors and for Nicholls, as the appraisal is encouraged to face problems head-on, and obviously disagreements arise. The Personnel Department's role in the process is one of advising and mediating. There is the opportunity to re-adjust gradings at Christmas of each year, but this is only used when obvious injustices require redressing. The appraisal system is designed to feed into and inform training plans.

The appraisal and grading system requires a considerable effort on the part of senior staff to understand their subordinates personally and professionally, and to communicate with them effectively. The preparedness to make such efforts is said to be an important concern when promotion is considered. A lack of effort in this field is viewed unfavourably at Annual Reviews.

Training

The company attempts to provide considerable training support to enhance career development. In 1989 Celltech staff attended 1350 man-days of training. The company's actual training budget includes resources for conference attendence. Over the past four years this budget has increased in absolute terms, but has declined as a percentage of company sales. The budget for training should be assessed, however, in the context of a company making little or no profit. Details are provided in Table 7.

Table 7. Celltech's Training Budget

	Training Budget £000	Training Budget/Sales %
1985	106	2.7
1986	123	1.6
1987	173	1.5
1988	193	1.1

The company's active approach to training is shown by its development in 1989 of a 13-week course in biotechnology jointly with Thames Valley College. It is based in the College's laboratories and was designed to train school-leavers in the basics of biotechnology. Celltech personnel were involved in the design of the course, although all the teaching is done by College lecturers. This course is believed to have been very successful and the quality of instruction to have been excellent. It cost £30,000, £17,000 of which was funded by the Training Authority. In November 1989 Celltech received a National Training Award from the Government's Training Agency in recognition of the innovative nature of this training scheme.

Celltech is deeply involved in graduate and postgraduate training. Apart from the company's 27 Science and Engineering Research Council CASE Studentships in 1989, there are five people studying for PhDs while working for Celltech.

Recruitment

Until recently, recruitment at Celltech has been almost entirely of graduates. It previously relied on informal links with the universities to attract these graduates. Nowadays, however, it more systematically does the 'milk round' at British universities and since 1988 has attended a number of recruitment fairs. The bulk of recruitment at Celltech is now amongst the lower graduate grades, reflecting the policy of 'bringing-up' raw graduates into Celltech's ways of doing things. Advertisements for more senior recruits are usually placed in journals such as New Scientist and Nature in order to reach an international audience. In 1988 the company recruited employees from the USA and France and is likely to increase the amount of international recruitment in the future. Headhunting companies are used occasionally.

Although in the past most of the activities of the firm, including technician-type tasks, have been performed by graduates, as the company has grown, built up the manufacturing function, and created more routine jobs, there has been a need to recruit lower grade school-leavers of 'A' level standard. A recent recruitment of 20 such school-leavers was the first significant batch of non-graduates to join the firm. The majority of these people will be working in manufacturing, although some may eventually work in R&D.

This recent recruitment drive had its problems and has required a novel solution. In the local labour market, most 'A' level school-leavers go on to further education. The company attempted to recruit in the high-unemployment North East and South West of England without success, and it eventually attracted the bulk of these recruits from Ireland, around Cork and Dublin. It has leased a company house to be used as lodgings for these employees.

Peter Nicholls describes the major problem of recruitment in Celltech as finding people sufficiently broad-based to match their scientific and technological skills with the personality to work, and develop greater future responsibility, within a commercial environment. In a fast growing firm, recruitment obviously has to consider not only immediate competences, but also potential future capabilities. One of the attractions Celltech might hold for recruits is that its fast growth offers promotion opportunities. The company does, however, face a number of difficulties recruiting scientific staff: particularly in the protein biochemistry and biology areas. Such personnel are in limited supply (Waite et al., 1988) and the problems of attracting them to Celltech are compounded by the high cost of housing in the locality; a particular disincentive to younger potential staff.

Celltech's creative use of both training and recruiting is revealed in the example of the establishment of the clinical/regulatory function. Staff in Clinical and Regulatory Affairs wereboth recruited in order to bring new skills into the company, and Celltech staff have been trained to develop new skills upon their existing competences. So, whereas four or five people were recruited on the basis of their extensive working experience in the pharmaceutical industry, others were trained and have developed their skills within Celltech. An example of the latter includes Dr Kathy Lambert who was a Celltech scientist and currently leads the Regulatory Affairs department. A balance has been achieved between recruiting new staff and developing new skills in existing staff (with the advantages that brings for individual career development and for the new department being able to use their knowledge of the company).

Organisational Structures

Celltech is a knowledge-intensive company – labour costs in 1988 were 38 per cent of sales [3] – and good communications and information flow are essential to its effective operation. A major facet of Celltech's strategy towards its employees is the encouragement of communications within the company. Gerard Fairtlough has a number of distinctive and strongly-held views on the importance of good organisation and communications as a mechanism for improving the efficiency of innovative firms. [4] Key amongst these is the concept of 'compartmentation', a term from cellular biology which describes the partial closure from the environment which allows multi-order feedback to develop between the components of a system. Fairtlough describes a number of features of cells attractive to organisational structures in firms:

– they are largely self-contained and are densely organised for intensive internal communication

– at the same time, they constantly communicate with other cells in a complex networking system

– despite their diversity, all cells live by rigid rules of competition and cooperation set by the larger cellular framework

– cells are continually evolving. New molecules emerge, new reactions are attempted, new relationships are assessed and either adopted or discarded. In such trial and error, however, only those relationships that promote the well-being of both the cell and the host organism are permanently integrated into the cellular life (Fairtlough, 1986).

Analogising from cells to organisational structures, Fairtlough considers three areas to be important when considering the best mechanisms for inducing good internal communications within a compartment. The first is architecture – the physical environment of buildings and layout, which can be structured to assist communications. The second is climate – the ethos and morale of an organisation. A climate conducive to innovation can be attained by openness and respect for individuals (these aspects are considered to require the greatest effort to inaugurate and maintain). The third is systems – methods for consultation and direction of effort. These are more readily designed effectively, and accepted, if the organisation's architecture and climate are correct. [5]

Continuing the cellular analogy, cells do not exist in isolation, and communicate intensively, and Fairtlough believes organisational compartments should do the same. Compartments should put great effort into spanning their boundaries, actively seeking knowledge from outside. Also important for organisations as well as for cells is awareness of the contribution of compartments to the greater whole. Compartments' contributions have to be validated and, if found lacking, may be discontinued.

It is interesting to assess how Fairtlough's cell structure simile accords with practices within Celltech. Chapter 8 will consider the extensive ways that Celltech links with its external scientific environment. In this respect it will be shown to have excellent boundary spanning skills. Some of the systems ideas have already been examined, particularly through the matrix form of the project management system. What follows is some consideration of the ways that personnel policies reflect attempts to create excellent communications.

These policies are realised in practice through a mixture of formal and informal mechanisms. The formal (systems) mechanism for employee representation in Celltech is a Staff Council. This body includes 14 nominated representatives from various geographical areas within the company (which now operates in five buildings). In 1986 the Staff Council was asked to propose ways of keeping good communication between management and staff as the company grew and to consider a representative attending Management Committee meetings. The view of the Staff Council was that there was little point attending the Management meetings, as anything of concern to the Staff Council would inevitably be communicated to them through other channels. Peter Nicholls believes the Staff Council to be an effective means of communication and representation, and that it has been responsible for introducing a number of changes in the company.

One particular facility designed to improve communications within the company lies in its new R&D building (architecture). In this building is an extended meeting area, where staff can congregate at any time of day. It always appears well used. [6] The company cafeteria in the same building is another venue for encouraging communications. It is always well attended (which is unsurprising given the excellence of the food and the high level of subsidy). A variety of other communications methods are used. All new company recruits are introduced to the 'strategy group' of senior managers at a drinks party, and there are often informal dinners organised following the completion of training courses. Dinners are arranged for managers Grade 8 and above to which their wives or husbands are invited.

The company has an extensive seminar programme, with both internal and external speakers. During the course of a year there are around 50 seminars given by Celltech staff and another 50 are given by invited external experts (boundary spanning). There are also a range of intra- and inter-functional discussions and meetings. In the Manufacturing function, for example, there is a weekly managers' meeting and a monthly meeting for all Manufacturing staff. The latter is frequently well-attended, and may on occasion include a presentation from someone from another function, for example, on a new R&D project. Additionally, there are monthly liaison meetings between Manufacturing and Development, and Manufacturing and Sales and Marketing.

In addition to these formal mechanisms, there is a great amount of informal communication within the company. Senior managers spend a significant proportion of their time talking to managers, scientists and other staff. To an extent this is formalised, for example, in 'management lunches' organised monthly by executive directors for all management staff, but there are numerous informal opportunities for discussions. The practices whereby managers devote considerable effort to knowing their staff are seen to be genuinely believed in, and effectively used, by the senior Directors.

Celltech is an openly-managed company (climate). [7] Some of the ways this openness has been manifested have, however, been curtailed as the company has grown. In the early years of the company, for example, a correspondence file, including all letters into the company not of a personal nature, was left in a room for all staff to read. The file became too bulky, and this practice was stopped. Also in the early days the minutes of Board meetings were left in the library. This has also stopped. Apparently some Board members were surprised to learn that this had been going on without their knowledge, and it was not long before it was extracts of Board minutes that were placed in the library. Nevertheless, the chief executive continues to report back to all staff on meetings of the Board. These company meetings are always well-attended and are perhaps indicative of the high level of information flow in the company.

Systems thinking has underlain much of Celltech's development (Fairtlough, 1989). The project management system is a revealing example of the way systems thinking is used in Celltech. The project management system is intended to assist the company with three key tasks: the generation of ideas for new products and technologies, the selection of the most promising of these, and their organisation for further development. Two typologies emerged within the company to deal with these tasks. The first classified the stages of evolution from ideas to products, the second classified the people responsible for organising each new stage (see Table 8).

Table 8. Project Management Typologies

Stage	Person Responsible
Idea	(Self Appointed) Idea Champion*
Candidate	Candidate Champion*
Research Project	Research Project Manager
Development Project	Development Project Manager
Product	Product Manager

*Part-time responsibilities

These typologies proved not to be quite right and they were amended. First, the idea of "Decision Gates" operated by decision makers were introduced. Ideas had to pass through these gates in order to become candidates; candidates had to do so to become research projects, and so on. Second to help decision makers reach the right decisions, two systems were introduced:

– A resource management system, which appraises resources and prioritises between projects.

– A project management system, which ensures that trained and experienced management were available for all stages.

The early stages depend on individuals, strongly committed to their ideas. In these stages the role of project management is to provide professional backing to the idea champions to make the best possible case to present at the next decision gate. This system is shown in Figure 13. (Fairtlough, 1989).

Such systems thinking has resulted in practice in the Research and Development functions being organised on a matrix basis with project managers coordinating resources drawn from the scientific departments. This is shown diagramatically in Figure 14. Project managers are responsible for budget approval, planning and control, establishing objectives and ensuring these are met. This enables the Heads of Departments in R&D to concentrate on keeping the resources of their departments (people, equipment and facilities) up to the required level and to maintain high standards of work practice and the scientific development of their group. He or she alone is responsible for the costs and revenues of projects. Nine project managers currently run 20-odd projects, the largest of which employs 17 people. They report to their respective Head Project Manager in Research and Development. Project planning for the larger projects is aided by computer.

The evolution of the matrix organisation provides insights into the thoughtful management decisions made at Celltech. In its early years, Gerard Fairtlough and Norman Carey spent a considerable amount of time thinking about project management systems which best met users' needs. (It will be recalled that both

Figure 13

Decision Gates

Decision Gates in the Evolution of Ideas to Products

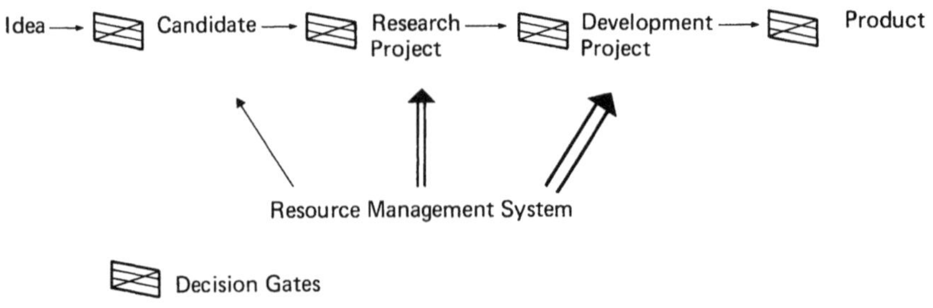

had extensive experience of working for large corporations, and were acutely aware of the advantages of organisational flexibility.) However the staff they recruited were scientists and generally unappreciative of the need for strong project management. The system eventually adopted – very similar to that originally envisaged by Fairtlough and Carey – was, therefore, not pushed on the scientists from above, but gradually evolved within the company as awareness grew of the problems department heads were having being both line managers and project heads. The present project management system therefore grew organically from the needs of the R&D function, informed and encouraged by top management. Because the system emerged in line with the broad realisation of the problems caused by previous structures, the transition to matrix management progressed without too many problems caused by conflicting authority structures. The system is believed to be working well, although it is known that this depends on the project managers working well with individual line managers and being committed to the matrix principle. For further discussion of Celltech's organisational practices see Fairtlough (1989). They also provide the basis of Dodgson (1989a).

The general principle used in organising the R&D function has been one of recruiting people with the requisite skills and then pulling them together in project teams. This has been used from the start of the company, and is preferred to the alternative of recruiting people for specific projects. The reasons for adopting this skills-based approach are, according to Dr Carey, that it: improves scientists' feelings of security (i.e. they feel they are needed for their skills, not just for a short-term project); and it encourages cohesive, problem-solving and creative activities, with people improving their long-term scientific skills as well

Figure 14

R&D Matrix Organisation

Department	Project 1	Project 2	Project 3
Molecular Genetics			
Molecular Immunology			
Microbial Systems			
Chemistry			
Protein Chemistry			
Protein Biochemistry			
Cell Biology			
Immunobiology			
Pharmacology			
Purification			
Process Development			
Fermentation Physiology			
Fermentation Technology			

as project skills. Skills are developed in the various departments and applied in the actual projects.

A major method for ensuring scientific excellence within R&D was introduced in 1988 with the creation of the Principal Scientist grade of employee. This innovation was brought in to allow appropriate financial reward for those people wishing to concentrate on developing their scientific expertise, rather than undertaking management responsibilities. [8] The Principal Scientists are appointed following an assessment which includes the use of external academic referees, and they are expected to have, and build upon, international scientific reputations. One of the Principal Scientists, Dr Emtage, describes this mechanism as an 'attractive career path'. He argues that the mutual concerns of Principal Scientists and line managers ensure that the demarcation of their activities is not clear-cut, and that the development of good working relationships between the two depends on the building up of trust and respect between the individuals concerned.

One of the mechanisms that Celltech has used to stimulate creativity has been the sanctioning of scientists spending up to 10 per cent of their time on their own projects. In its earlier years this was an explicit policy, whereas currently the practice is less formally constructed. It has always been necessary for those wishing to undertake these so-called 'preliminary studies' to have the approval of the Head of Department. It is still possible for scientists to undertake these studies, and this may on occasion exceed 10 per cent of their time. The intention underlying these arrangements is to allow scientists flexibility to use their knowledge in a direction unconstrained by project requirements. It is believed that it is important for the company to provide enough time in the face of immediate commercial demands to allow scientists to develop their know-how useful for a wider number of projects than their immediate one. In practice, this has been difficult to arrange, and there have been tensions between short- and long-term pressures.

Conclusions

The company's personnel strategy diverges considerably from that in traditional UK industrial firms. It attempts to recreate features of what its chief executive views as favourable characteristics of cellular communications systems. Very importantly, the senior directors are committed to, and have created a climate of, good personnel management and internal communications, and make efforts to infuse this enthusiasm throughout the firm. This is considered essential if the firm is to retain the benefits of small size as it grows.

Traditional industrial relations concepts such as 'discipline' and 'control' are essentially meaningless within the company, as work-pace is more determined by individual motivation and peer pressure than by organisational sanction. The company's 'high trust' attitudes to its employees is described in this quotation from Gerard Fairtlough.

"In order to attract, retain and motivate highly sought after people and to generate activities with high creativity and productivity it is essential to treat everyone in the Company as a responsible adult with a stake in the success of the Company; to share information (sometimes highly confidential information) widely within the Company; and to develop everyone's personal potential to the full (even if this sometimes means that people are enabled to get attractive jobs outside the Company." [9]

In Celltech, the major efforts of the Personnel function are recruitment, training, assistance with individuals' appraisal and the provision of incentives and career development. While the culture of commitment to personnel matters has existed in Celltech for some time – and is probably inevitable in a knowledge-intensive firm – the mechanics of practically running the system appear in the past to have been under-managed. Recent efforts have been made to systematise many aspects of the Personnel function: personal records have been computerised, graduate recruitment is approached more systematically. With the recent influx of new staff, particularly in lower grades, some new non-scientific training courses have been devised and have apparently worked successfully.

One of the major reasons for the increased 'formalisation' of the personnel system has been Celltech's increased size. How the company can retain its flexible personnel system as it grows is an open question (although it is being seriously addressed within the company). For example, the grading and appraisal system offers considerable flexibility in, and potential for, the provision of incentives, for the creation and maintenance of consensus, and as a means of communication. It is, however, enormously time consuming, both for the Personnel Department and for individual appraisors. With the Personnel Department already finding the volume of work and the pace of change its major problem, questions arise as to how this system can be maintained if the company grows much larger. While there are inevitable learning effects, as appraisors and appraisees become more accustomed to the system, the volume of new recruits may inhibit its effective use if the company continues to grow.

Continuing growth may also affect the 'climate' of the company. Close attention to the needs of individuals (and close proximity) is a major mechanism by which motivation is engendered within the company. Celltech aims to ensure that, as it grows, it attempts to maintain the high levels of commitment and motivation shown within the company. This motivation is very important to the company's competitiveness and its ability to do things faster or more efficiently than its competitors. The high motivation shown by Celltech employees has in the past manifested itself in extraordinary efforts. One manufacturing manager, for example, explained how, in order to save a batch of a product when a boiler broke down, he worked for 48 straight hours over a weekend to keep the process working. That this 'hard work' ethic continues is shown by a number of anecdotal instances, such as the high number of cars remaining in the car park after 6.00 pm, and a manager explaining to the researcher his guilt at having to leave the office at 5.30 pm, knowing his staff will remain for at least another hour. One company director told the researcher that his last task every day was at 7.00 pm to insist to his staff that they go home.

Celltech's very high emphasis on organisational and human resource issues is also found in other DBFs. To quote from the President and chief executive officer of Cetus:

"... how do we compete with them (large firms like Merck and Lilly) now that they are in biotechnology too? We do it by attracting, motivating, and retaining the best and the brightest, giving them a piece of the business through stock options and providing them with an environment where they can express their creativity in a way that cannot as yet be found in the big companies...The challenge for the manager is nurturing this potent yet fragile environment during a period of rapid change and growth". (Fildes, 1990).

Footnotes

[1] Morgan and Sayer's (1989) study of electronics firms in the M4 corridor (of which Slough is a part), reveal a number of features of the industrial relations systems in these firms similar to Celltech's. These include emphasis on training, intensive screening of recruits, encouragement of self-motivation, and a flexible payment system based on individual assessment.

[2] There are 13 grades (the Executive Directors are not graded). By way of illustration:
Grade 13 and 12: Directors.
Grade 11: Senior Management.
Grades 10 and 9: Middle Managers.
Grade 8: First Manager level.
Grade 4: Newly appointed PhD.
Grade 3: Newly appointed graduates.

[3] By comparison, labour costs as a percentage of sales in Glaxo were 20 per cent; Beecham, 22 per cent; and Wellcome, 27 per cent (1987 figures).

[4] See, for example, Fairtlough, G: 'Creative Compartments', London Business School Journal, Summer 1986; and 'Can We Plan for New Technology?', Long Range Planning, Vol.17, No.3, 1984.

[5] 'The principle of compartmentation requires small, self-contained, self-managing units. Compartmentation is a design principle aimed at ensuring that knowledgeable people interact as easily and creatively as possible. In a knowledge-intensive society it is critically important. Physical or architectural arrangements must help people to get together. Systems and procedures must meet the needs of operating people,

rather than those of systems designers or just of the management. Above all, psychological barriers to communication must be removed as well as physical or systems barriers. This means an open style of management is essential. The intense interaction within a compartment which develops when the architecture, systems and climate are right leads to a creative, problem solving, self motivating organisation.' Fairtlough, G, Society Medal Address, 'Getting Organised for the 1990s', Chemistry and Industry, 19 August 1985.

[6] It is perhaps indicative of the ethos of the company that when I once asked a director if this facility was abused, I was told my question was meaningless. The facility was there to be used, the more the better.

[7] Perhaps one of the greatest examples of openness within the company is this book. The author has been given complete access to information and staff within the company. During the course of the research I have continually been struck by how much information managers, scientists and other staff have been prepared to offer me. It has also been refreshing to find that this information has been given 'warts and all'; mistakes have been admitted. A number of long-serving Celltech employees have made the point that although the company is not as open as it once was, it is still very open in its managerial style.

[8] Principal Scientists are graded from Grade 8 (equivalent to Head of Department) to Grade 12 (Divisional Director).

[9] Personal communication to author, 26th February, 1990.

Chapter 7
Technological Learning

Introduction

Science and technology provides the basis of Celltech's past, present and future competitiveness. This chapter will describe and analyse Celltech's accumulation of biotechnology know-how. It will focus on the strategic acquisition of technology and the question of technological learning. The competitive position of DBFs like Celltech has depended first, on building up exceptional R&D skills based initially on academic links, and second, on developing an integrated business capacity with distinctive advantages over large pharmaceutical firms based on their novel technological skills and organisational flexibility. These commercial strategies depend on the company being able rapidly to acquire and diffuse technological knowledge, both theoretical and practical. Acquisition of such knowledge has an internal component – which will be the focus of this chapter – and an external component – which will be concern of the following chapter.

The Evolution of Celltech's Science and Technology

This section will describe the evolution of the main strands of Celltech's science and technology. It will not describe these strands in detail. [1] Basic descriptions of MAbs and rDNA have already been provided in chapter 2, and some of Celltech's expertise in both these fields has been described in chapters 4 and 5. A number of observations about Celltech's technologies can be made from the outset by referring to Figures 15 and 16. Figure 15 shows the approximate periods when Research and/or Development were active in the indicated technologies. Periods of significantly increasing or decreasing activity are indicated, but the Figure does not signify relative actual resources deployed. Dotted lines indicate continued low level activity, but not active development of systems/technologies. This Figure shows that Celltech remains active in 1990 in all the technologies in which it was active prior to 1983. With the exception of E.Coli expression systems and MAb fermentation, all the technologies during this period have remained equally or increasingly important to the company.

Celltech's decision to develop a broad initial technology base can therefore be seen, in the light of its subsequent strategy to become a pharmaceutical company, to be well founded, as the selected technologies continue to remain important to the company with this aim. Figure 16 shows another benefit of Celltech's original strategy. It portrays schematically the convergence that has occurred between MAb and rDNA expertise.

Essentially, Celltech's early expertise was derived from external sources. On the MAbs side involvement began with its early production of antibodies to alpha-interferon (derived from the MRC), and blood grouping reagents (also derived from the MRC). Its rDNA involvement began with the considerable intellectual capital recumbent in scientists recruited from GD Searle. It was not long, however, before the company began to develop its own proprietory, self-generated expertise. Mention has already been made of the way that Celltech brought in from the MRC a replica of the crude method for manufacturing blood grouping reagents, and the efforts to improve it. After comparing the various possible types of equipment which might be used for large-scale mammalian cell culture, John Birch's team chose an airlift fermenter as the most suitable means for cells such as hybridomas to grow in suspension. Thereafter, Celltech was breaking new technological ground; it was attempting to do something that previously could not be done and carried it further than other firms.

A significant process engineering effort was put into the detailed design and choice of materials and equipment, and into the development of automation for

Figure 15

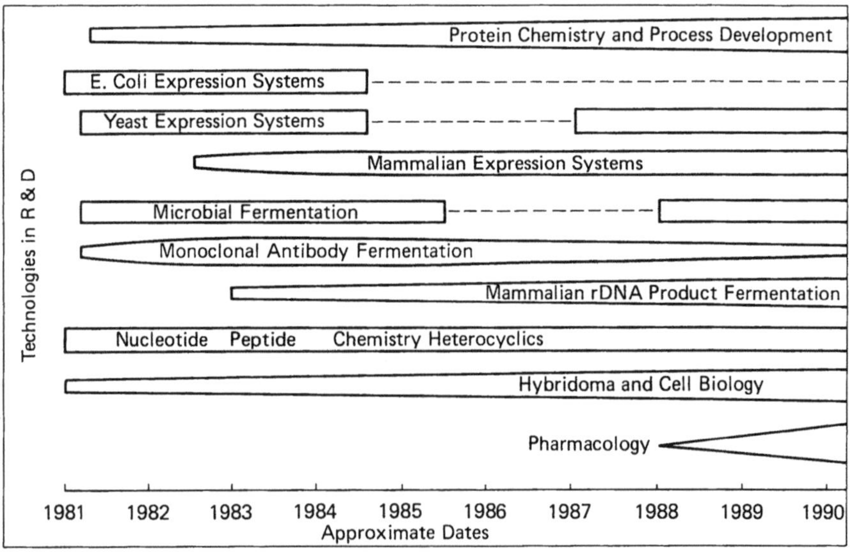

cell culture plants. Extensive work on cell physiology, applying many of the approaches used in microbial physiology to mammalian cells, and then in process optimisation, went on alongside the process engineering studies. [2] Celltech's novel development in this field was a considerable technological feat, and provided the basis for the company's move into bulk production of biotechnology products. [3] Its 'cumulative learning' in this area has continued to be of advantage.

On the rDNA side, we have seen how Celltech's work on chymosin, led by an ex-GD Searle scientist, enabled the development of proprietory technology for the expression of cloned genes. This proved particularly important when the company began to extend this expertise to the study of the expression of antibody genes. [4]

In its early years Celltech spent a good deal of time establishing its basic expression systems. It examined three main methods of expression: E. Coli, yeast and mammalian cells. The first of these, Escherichia Coli, is a fairly simple bacterium and a well-known system in which to express proteins. This system took around three or four years to master. One system was considered insufficient, and an extensive collaboration was established with researchers at Oxford University to work on yeast. This proved unsatisfactory for a variety of non-scientific reasons. Three or four mammalian cell systems were chosen for study from the many available. All were eventually used, but the company finally selected chinese hampster ovary (CHO) and myeloma as its primary cell

Figure 16

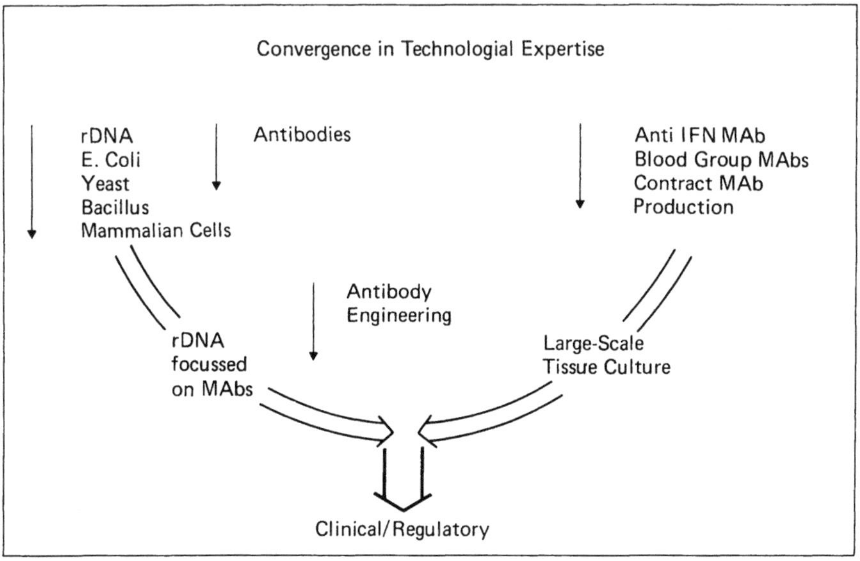

systems. [5] Apart from its innovative work on using mammalian cell for expression, the company has also developed some innovative systems in yeast and in micro-organisms, such as B. Subtilis.

Once it was established that a company working in both MAb and rDNA technology could conceivably combine the two, a research team directed by Dr Spencer Emtage was created in order to learn how to clone antibody genes and express them in various cell types. This, again, was Celltech attempting to do something new. Furthermore, it was something that had no targeted products, and was very much exploratory research. The bulk of the leading research into so-called 'antibody engineering' was (and is) being undertaken at the LMB at Cambridge. [6] At this time the MRC undertook no patenting itself, and Celltech was in a position to capitalise on some of this work by assisting the MRC with its patenting. As we shall see, however, some arrangements proved very difficult to negotiate. Nevertheless through its own efforts, and through links with the LMB, Celltech did develop expertise in antibody engineering. [7] Scientific papers based on this technology began to emerge from Celltech scientists in 1984 and 1985, and at the same time the company filed some related patent applications, which it claims are the earliest in the world of this type. Antibody engineering highlights the company's ability to assimilate and develop externally-derived technologies.

The consolidation of Celltech's expertise in the two technologies of large-scale manufacture of proteins in cell culture and antibody engineering, and their convergence, has provided the company with its core technological competences, and the basis of its competitive position. It enabled the company to embark on its antibody-based drug discovery programme. The latter has, of course, involved the company developing new areas of scientific expertise, particularly related to clinical and regulatory issues. This has resulted, for example, in the appointment of a medical director, Dr Alastair Riddell, in early 1987, and the establishment of a pharmacology department.

The creation of the clinical function and of pharmacological expertise over the past few years has been a major development within Celltech. When Dr Riddell joined the company it had no products in clinical trials, and the clinical and regulatory function employed only two people. It currently has three products in clinical trials (anti TNF, CGRP and Theramabs) for a wide range of illnesses and employs 18 people. The new skills developed in the company are organised into three departments: Regulatory Affairs (which is believed in the company to be in the area the most competent group in Europe); Clinical Research Group, for the monitoring of trials; and a Data and Statistics Group.

Celltech's Technological Progress

Measuring a company's technological progress is fraught with difficulty. Any chosen indicator can only be partially meaningful and can only be considered as a proxy measure. The most commonly used measures are, on the input side, R&D expenditure, and on the output side, patents granted. R&D spend might reflect commitment to technological development, but it does not, of course, reveal the effectiveness of any expenditure. Celltech's R&D expenditure has already been considered in Chapter 2, and has been shown to be extensive: in 1989 total R&D spend was equivalent to 51 per cent of company turnover. Figure 6 shows that the major proportion of R&D undertaken is from Celltech's own account: some £6.2 million in 1989. It would appear that the company's increasing R&D commitment is matched by increasing patenting activity. Figures 17, 18 and 19 provide details on the extent and nature of Celltech's patent portfolio, and show the significant increase in the number of patent applications made in each year of the company's existence.

It appears from the limited data available, that Celltech, on the basis of patent applications made, is a significant contributor to British industrial biotechnology R&D output. It is, however, very difficult to assess Celltech's patent position. There is a time lag between patent applications and their being granted, which,

Figure 17

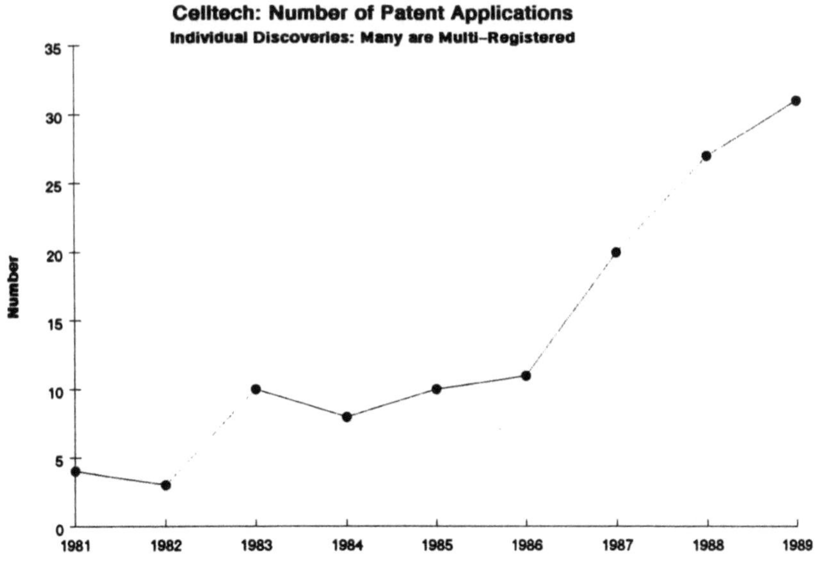

given the recent nature of the company's patenting efforts, makes it perhaps more meaningful to consider applications rather than patents granted. The data shown in the figures relate to 115 patent applications which are still current, i.e.

Figure 18

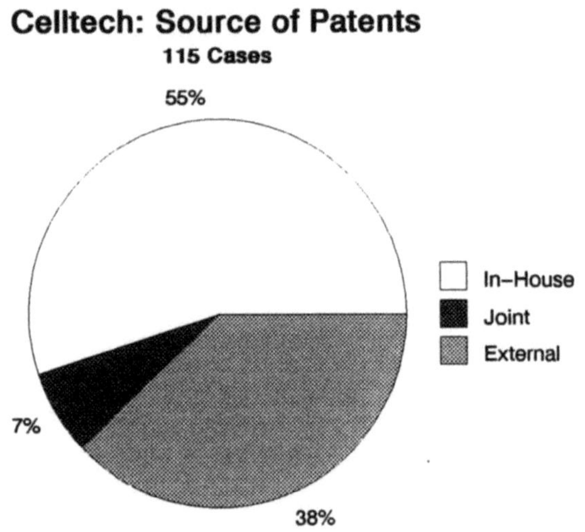

Figure 19

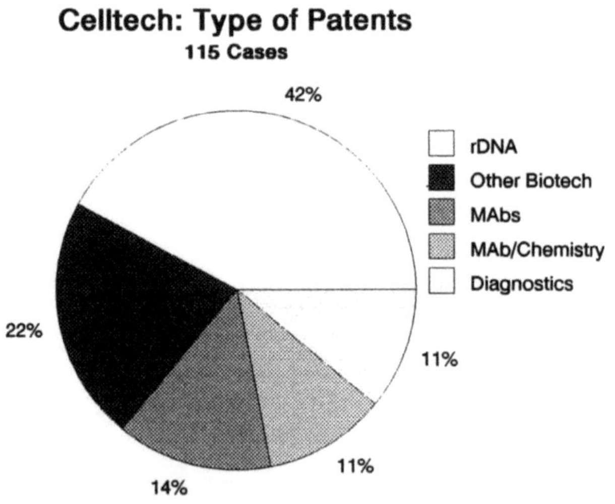

a number of applications which have been withdrawn are not included. The majority of patent applications have been made in more than one country, so the data are for separate discoveries and are not the number of filings.

To put Celltech's patenting performance into context, data derived from a biotechnology patent application database: Biotechnology Abstracts, is used. This data records all biotechnology patents filed and includes worldwide applications. Table 9 provides data on total applications to November 1989 by some selected companies.

Table 9. Biotechnology Patent Applications by Britain's Leading Pharmaceutical Companies

Company	No.
ICI	79
Beecham	47
Celltech	44
Wellcome	34
Glaxo	19

Source: Biotechnology Abstracts

Searches were also made for the two highest-profile British pharmaceutical biotechnology companies, apart from Celltech: British Bio-technology and Porton International. British Bio-technology recorded no applications and Porton Products recorded two.

By way of comparison with US DBFs patenting performance, Cetus, the longest-established DBF had 90 US patents awarded and 199 awarded elsewhere, and it had 296 US patents pending with 1900 patents pending elsewhere. Chiron, a company of a similar age to Celltech, had 8 US issued patents with 7 elsewhere, and 59 US patents pending and 124 elsewhere. Figures are for 1988, and are derived from the company 10K reports.

There are numerous problems with using this patent data authoritatively. Not least because of the discrepancies between Celltech's recorded patent applications, and the numbers recorded on the database. Assuming that there is a similar discrepancy for all the companies involved, then Table 9 does tend to show that in terms of patent applications, Celltech is equal to some of Britain's leading pharmaceutical companies. However, numerous caveats must be placed on the interpretation of patenting data, not least relating to the fact that numbers of patents do not reflect their quality.

In the first five years of Celltech's existence, slightly more of the 25 patent applications made came from external sources, such as the MRC and universities, rather than being based on internal R&D (15 external to 12 internal). Since 1985 many more applications based on internal R&D have been made (51 internal to 29 external). Greater proportions of the company's acquired knowledge therefore appears to be derived internally. The small numbers involved limit the value of considering any trends in the nature of patent applications. However, whilst applications in MAbs and rDNA have been a feature since the company started, it is only since 1986 that applications have begun to emerge from work on the chemistry of MAbs.

One other measure of Celltech's technological progress can be suggested, specific to the nature of the company's technology. This is the production capacity from the company's fermenters. Bulk production of biopharmaceutical products is an extraordinarily complex process, with contamination a permanent threat. Celltech has pioneered much of the development of this technology, and therefore its expanding capacity may be used as a proxy for technological achievement. Figure 20 shows diagrammatically how rapidly this capacity has grown. In 1984, the company had just over a 1300 litre capacity; this increased to 2500 litres in 1986, and to 4700 in 1988. [8]

Learning by Making

The ways in which Celltech attempts to encourage learning from R&D has been described in previous chapters. We have seen the considerable extent of

Figure 20

Celltech: Cumulative Growth in Manufacturing Capacity

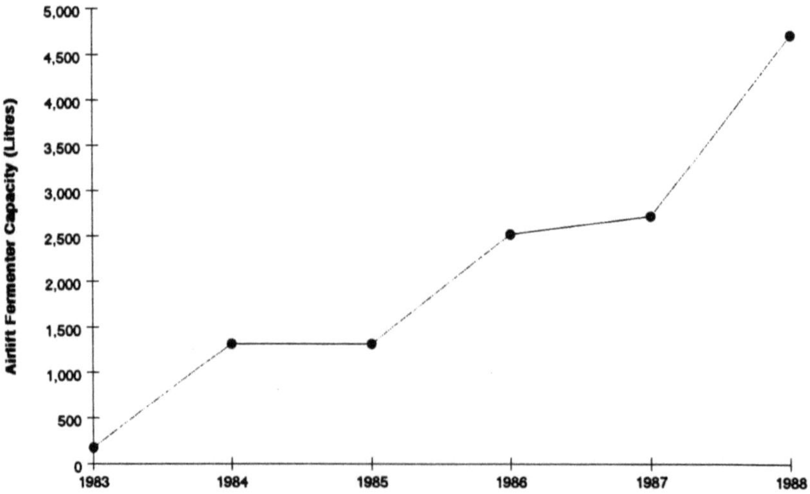

Celltech's expenditure on R&D, both own-funded and sponsored. This commitment of financial resources is matched by human resources. The Research function, for example, currently employs around 100: around 30 in the Chemistry and Biology sub-divisions, and 40 in Molecular Genetics. The previous chapter described how the company attempts to encourage the creativity of R&D and its receptiveness to external sources of knowledge. What follows is a description of the ways Celltech learns by manufacturing.

As we have seen, Celltech's world-leading manufacturing capacity has developed rapidly, and has involved the company creating new production techniques. Two examples from the Manufacturing Division highlight the importance of another aspect of technological learning, that of the cumulative extension of in-house expertise. The first is in fermentation, the second in roller bottle processes. The Manufacturing Division has grown rapidly since 1986; it then employed 35 (including Quality Assurance), and it presently employs 118 (excluding Quality Assurance). The expansion of Celltech's fermentation capacity is shown in Figure 20. The original airlift fermenter was purchased in 1983 and proved to be of poor design and quality: it was supposed to have a 100 litre capacity, but could actually only manage 80 litres. Later that year another 100 litre fermenter was bought and this produced the whole 100 litres. In 1984, both these fermenters were re-designed and modified, and both managed to produce 110 litres. The first 1000 litre fermenter was commissioned in 1984, some eight months after it had originally been planned. Another 1000 litre fermenter was introduced in 1986. This was fabricated entirely off-site, and was placed in the company's purpose-built manufacturing building. This later 1000 litre fermenter took around four or five months to commission.

There were major technical problems with these fermenters, including the 1000 litre fermenter introduced into the new production building. When the Manufacturing Director, Mike Harvey, was recruited in December 1986, he inherited a high failure rate in batches (up to one in three failed), caused by microbial infections. These failures were exacerbated by plant teething problems and some major failures occurred in the new building, such as the boiler breaking down. To deal with these contamination problems, Harvey began to examine the whole process systematically. He found that the plant was perhaps only 95 per cent right in its design, and needed modification. A multidisciplinary team was established to look at these failures, and a microbiological survey was undertaken on how to improve the system. Eventually the contamination problems were eradicated, and while there are occasional production problems, these have more to do with the complexity of the cells being used rather than equipment problems.

According to some in the company, the problems with the early manufacturing processes derived from insufficient investment and time to plan properly. Everything was rushed, and corners were cut. Given these circumstances, it is perhaps remarkable that Celltech has attained its leadership position in bulk production, particularly given the problems that other companies have had attempting to do the same. That these problems were overcome owes much to the dedication and problem-solving skills of the Division's employees. That the lessons from these experiences have been learned, and used effectively, can be seen in the example of the company's latest fermenter.

This latest fermenter, with a 2000 litre capacity, has been designed completely in-house. Planning for this piece of equipment began in Spring 1987, and it was commissioned a year later. Since its commissioning this fermenter has worked continually without major problems.

In 1987 Celltech received a contract to assist a subsidiary of Johnson and Johnson to manufacture EPO. EPO is manufactured by means of roller bottles (Essentially, bottles which are rotated and encourage the growth of product on the side of the bottle). During 1988 a production facility was established and assistance with the production of EPO commenced. The same year, Celltech was awarded a further contract to assist the manufacture of EPO, and began to build purpose-built facilities. This entire production plant, which was designed, and its building managed, by the company's Production Engineering Department, was operational a year to the day building work started. Due to this extraordinary effort, the plant currently produces three times more EPO than the original plant. The speed by which it became established provides testament to the breadth of skills currently within Celltech and its ability to build upon existing skills.

Pisano, Shan and Teece (1988) argue that new biotechnology firms 'lack the key input necessary for competitive large-scale bioprocessing, namely, production know-how gained through experience' (their emphasis). Celltech provides an exception to this rule, and has done so through its ability to learn quickly. Not every problem can be foreseen, of course. The new EPO plant is an impressive manufacturing facility. Apart from the arrays of special equipment and facilities, such as clean rooms and cold rooms, the plant also incorporates equipment generally not associated with biotechnology firms, such as some pick-and-place robots. The author visited the plant shortly after its commissioning, and obviously not every problem had been foreseen. There was no facility, for example, for packaging the hundreds of used bottles before their incineration. It befell an unfortunate young recruit to spend his days loading bin liners with bottles.

All these developments in the manufacturing facilities involve considerable efforts. In August 1988 the Board considered the 'extraordinary pressure the Manufacturing Division had been under'. It had faced contamination and technical problems, and had constructed an EPO manufacturing facility and a hGH plant, and had planned and commenced construction of the further EPO production facility. In addition it had been subjected to intensive inspections by the FDA, DHSS (Department of Health and Social Security) and Canadian regulatory officials. Some lost production through mistakes had occurred, but intensive training programmes for junior staff were aimed at alleviating these problems.

Celltech's development of manufacturing processes continues. Since the establishment of the EPO plant the company has enjoyed some success in developing a microcarrier based unit process for growing anchorage dependent cells as an alternative to the labour intensive roller-bottle process.

Footnotes

[1] The science and technology of biotechnology can appear baffling to the non-expert. Within Celltech a number of people have become skilled at explaining biotechnology to those with little biological or biochemical knowledge. For that reason I recommend those wishing to know about biotechnology in greater depth read: Fairtlough, G, 'Make Me a Molecule: the Technology of Recombinant-DNA', Proceeding of the Royal Institution, 24 February, 1984; and Carey, N, 'Production and Use of Therapeutic Agents', British Medical Journal, 10 October, 1987.

[2] Fairtlough, G, 'Exploitation of Biotechnology in a Smaller Company', Biotechnology – Spinks Eight Years On, Royal Society, 13/14 October 1988.

[3] During the course of my interviews with people inside and outside Celltech, I asked the question 'What has been Celltech's greatest achievement?'. By far the most common response has been John Birch's achievement with bulk manufacture. One very senior British scientist described Birch as a 'national treasure' for what he had achieved.

[4] Although Celltech originally pursued the two main strands of biotechnology, I have seen no evidence to suggest that this was undertaken in the view that they eventually could merge. To this extent the scientific convergence was fortuitous for the company. The realisation of the potentials provided by the convergence does, however, reflect the company's sensitive and adaptive strategy. According to Norman Carey, Celltech's early realisation of the opportunities this could provide was manifested prior to the appointment of Dr Mike Boss in January 1982, who was recruited specifically to fuse the two areas of expertise.

[5] All these expression systems – E. Coli, yeast and mammalian cells – allow for the controlled production of proteins in fermentation systems. The cell system chosen to express a protein depends on the nature of that protein and the use to which it will be put. For example, if a large quantity of the protein is required the choice would probably be a micro-organism as the means of production. If the protein is large and complex the choice is more likely to be an animal cell in culture (Carey, 1987). Although manufacture through mammalian cells provides the most advantageous means of producing complex proteins, mammalian cells themselves are complex and fragile and their use in production involves significant expertise.

[6] Gerard Fairtlough, in a speech to the Royal Society in 1988, argued that a number of scientists at the LMB, including Drs Neuberger, Winter and Rabbitts, laid the scientific foundations for the genetic engineering of antibody molecules. These

include discoveries related to size reduction, incorporation of specific attachment sites, the making of chimaeric molecules (mixed human and mouse regions) and the grafting of Complementarity Determining Regions (CDRs) from one antibody to another. (CDRs are the part of the molecule which gives it its specificity and to be able to retain the antigen specificity of a mouse antibody in a molecule which otherwise is a human antibody has enormous potential.)

[7] Dr Emtage describes antibody engineering as 'making antibodies better products'. Celltech is attempting to combine the best aspects of mouse, rat and human MAbs, in the belief that grafting CDRs from mouse antibodies into a human antibody framework, is more likely to to meet ethical, manufacturing and patient response objectives, and also be more patentable. The following table is taken from the company's 1987 Prospectus.

Table 10. The Relative Advantages of Types of MAb

MAb		Relative	Advantage	
	Ethics of Immunisation	Production Scale	Patient Response[1]	Patentability[2]
Mouse	+	+	–	?
Rat	+	+	–	?
Human	?	–	+	?
Human (Engineered)	+	+	+	+

[1] Products from mouse and rat cells suffer the disadvantage of provoking an immune response in humans which with repeated doses renders the MAb inactive.
[2] Humanised (engineered) MAbs are expected to be more readily patentable, since they are modifications of naturally occurring entities.

[8] Celltech receives universal plaudits for what it has achieved with bulk manufacture. In interviews, Dr Brian Richards, Chairman of British Bio-technology, said that 'Celltech's manufacturing facilities are as good as any in world. They run a quality operation in every sense of the word.' And Mr Mark Cantley, the Director of the European Community's biotechnology programme, describes Celltech as having a manufacturing capability as good as any in the world. Another external expert I consulted compared what Celltech had done with the conspicuous lack of success of a number of competitors attempting to do the same thing.

Chapter 8
Building External Networks

This chapter focuses on the range and intensity of relationships that Celltech has established with external organisations and which have proved very important to its acquisition of science and technology. It considers commercial technological linkages, links with academia and its centrally important collaboration with the MRC.

Commercial Technological Linkages

The development of biotechnology has been characterised by extensive collaborations between firms. [1] These have been a particular feature of biopharmaceutical development and have tended to focus on R&D. [2] Earlier chapters have described some of Celltech's contract collaborations with large multinational companies. [3] Celltech has obviously benefited from these collaborations financially and from the building up of skills and competences necessary to undertake the contracts. It has also been a strategy of the company to attempt to derive benefits additional to those related to the specific project.

Within Celltech three generic types of technology are conceptualised: background, enabling and project specific. These are schematically portrayed in Figure 21. The classification relates to the interactions between research which can be done for contract clients and the company's own-products and technological development. Background technology is all the technology existing in the company at any point in time which is not contractually allocated in some way. In contract terms, it is pre-existing technology belonging to Celltech before the start date of a new contract. When undertaking any contract, both project specific and enabling technology is generated. Project specific technology is directed towards, and is applicable only to, the objective of the project. It must be very tightly defined, and it will belong, or be licensed exclusively to, the project sponsor. Enabling technology is developed during the contract, but is applicable in wider fields than that which is the objective of the project. Celltech retains this technology but it would license it non-exclusively to the project sponsors insofar as it was necessary to practise the project specific technology. Enabling technology involves learning and it is one of the central purposes of contract research to build up enabling technology so that in subsequent projects it becomes background technology.

A prime practical example of this strategy is provided by the therapeutic MAbs contract for American Cyanamid. In this programme, designed to develop MAb-based aids to cancer diagnosis and treatment, Celltech has retained all rights to non-cancer discoveries. It has formalised its expectations regarding the development of enabling technology. In the past, however, building up enabling technology has often been a problem. A balance has had to be retained between meeting short-term (project) objectives, in order to earn income, and long-term objectives of building the company's science-base. In practice, managers have had to be extremely flexible in conflating these long- and short-term demands. [4]

Potentially this tension between long- and short-term needs could manifest itself in disputes between line managers and project managers. To circumvent these, and to ensure that enabling technology is given a high priority, project managers are not asked to deal with the allocation of resources between project and enabling technology. Such decisions are referred upwards in the organisation; the bigger the decision, the higher it is referred.

Academic Linkages

Links with academia have, of course, been very important for the company. Celltech's arrangements with the MRC have been crucial to its formation and central to its past and present scientific development: these will be examined

Figure 21

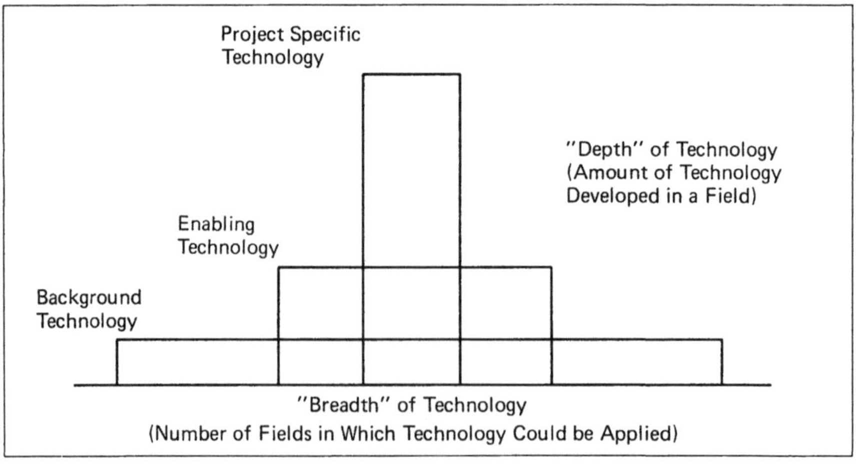

below. Another important mechanism for linking the company with academia, that has also existed since its formation, has been the Science Council, an academic consultative body. The initial rationale for the formation of the Science Council followed the US experience whereby it was primarily academics who set up DBFs. In the absence of this academic-driven impetus, Celltech was felt to need a strong academic input. In this sense, the Science Council did not grow from any intrinsic, specific need but more from the belief that there was a role for a consultative body which was integrated into the academic scientific community. Many US DBFs, such as Amgen and Cetus, have similar scientific advisory bodies.

The membership of the Science Council has evolved over time, as has its role. Initially it reported to the Board, but its then chairman, Sir Michael Stoker, changed this arrangement as he believed that it would better serve the company by reporting to the company's management, a move endorsed by John Jackson. Direct access to the Board was, however, also retained. The Science Council has always included very eminent scientists (including for a short period the Nobel prize-winner, Cesar Milstein). Its role is documented within the company as being:

– to bring new ideas to Celltech and to help in filtering new ideas coming from any source;

– to respond to requests from Celltech's R&D Management for reviews of major topics (e.g. large new project proposals);

– to advise on Celltech's general relationships with academia (e.g. styles of collaboration, levels of fees, recruitment methods, Celltech's PhD programme etc.);

– to create sub-groups (e.g. a clinical group);

– for very senior appointments, informally to draw up a specification for candidates, and in locating and appraising candidates.

In practice, because of the members' disparate backgrounds, unified proposals and recommendations have been rare, and the view is held in the company that the greater value from the Council is achieved through what it offers as a collection of individual consultants. [5] The Science Council used to meet six times a year and discuss a set agenda in plenary session. Now it meets less often, and in specific groups, as and when required.

Since its formation, Celltech has always had a senior manager responsible for academic liaison. At present, this activity is the responsibility of Dr Gwyn Humphreys, who as Director of Academic Liaison oversees contacts with universities, research laboratories, hospitals and some small firms.

In 1988 £1.25 million was spent by Celltech on its academic collaborations. Most of these collaborations tend to be longer term than simple licensing agreements. Typically they involve Celltech funding a laboratory or individual researchers (usually for three-year projects) in return for the assignment of exclusive rights of any discovery to Celltech. The company has, however, to be flexible in its negotiations, as it is impossible to control all intellectual property rights in scientific networks. Some of these collaborative projects are complex, and this is shown in Figure 22 which reveals one such network of coordinated research collaboration. In 1988 around 20 three-year collaborative and consultancy agreements were negotiated by Celltech. The reasons the company gives for its high level of academic sponsorship are as follows: [6]

– to seek products and ideas to complement Celltech's projects and programmes;

– to seek information and discoveries which will/can develop or alter strategy (awareness and serendipity);

– Celltech as a company provides a 'product path' for basic scientific discoveries, many (if not most) of which are made in academia;

– academia provides a resource for specialist (and expensive) technology, e.g. nuclear magnetic resonance;

– through some funding/sponsorship Celltech contributes to healthy survival of the academic base in the UK;

– PR activities aid/promote recruitment flow into Celltech;

– training Celltech employees in scientific techniques.

Building on its initial links with the MRC, the company has sought to access the broader base of British and foreign scientific expertise. Table 11 gives some indication of the extent and range of collaborations Celltech had undertaken by

Figure 22

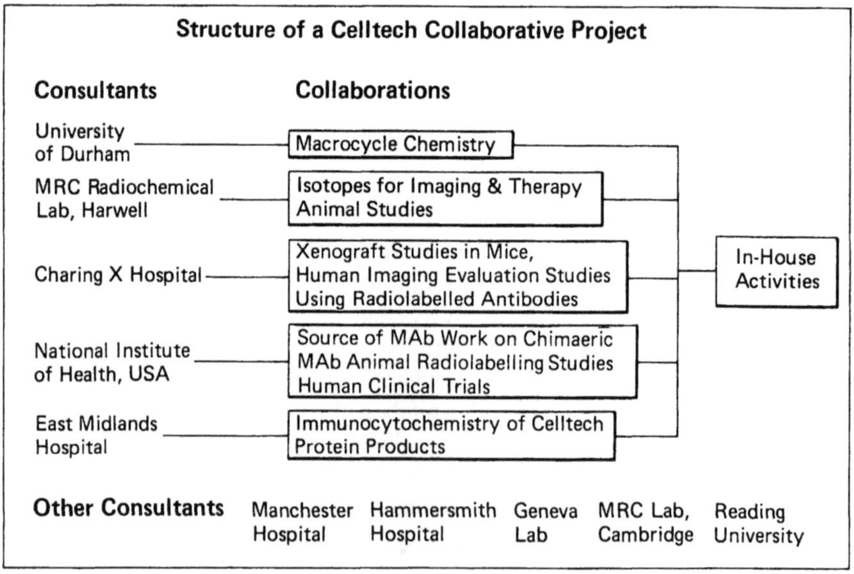

1989. It shows that over one-third of Celltech's collaborations have been with overseas partners.

Table 11. Collaboration with Universities and Research Institutes

Universities	Country	Research Institutes
19	UK	9
3	USA	3
2	Netherlands	1
1	Japan	1
1	Switzerland	
	Belgium	
	FRG	1
	France	1

The value to Celltech of academic collaborations can be seen by the extent of academic involvement in its major current and past activities. Significant amounts of the research for CGRP was conducted at Middlesex Hospital and the Royal Postgraduate Medical School. Anti-TNF involved research at the University of Maastricht. Much of the research into antibody engineering is undertaken at LMB. The research for hGH was undertaken at Georgetown University, Washington DC. Acid Stable Lipase started in collaboration with St Mary's Hospital Medical School.

In the year from October 1988, 55 per cent of Celltech's external R&D expenditure was related to what it calls 'near-term' products, such as CGRP; 11 per cent was related to mid-/long-term products; and the remainder was related to the company's technology base. The proportion spent on mid- to long-term products has been relatively stable over the last few years.

Celltech is involved in a variety of forms of academic liaison, including a number of government and research council research initiatives. Table 12 includes some details of the nature of some of Celltech's current collaborations.

As a proportion of Celltech's total academic collaborations, government and research council initiatives account for less than 12 per cent of expenditure. As a proportion of Celltech's entire R&D expenditure, its involvement in these public sector schemes accounts for only around 2 per cent.

Technology Watching

A variety of mechanisms have been described by which Celltech's scientists are kept up-to-date on developments in their field. To this list can be added the

Table 12. Celltech's Academic Collaborations[7]
(Current October 1989)

Research Collaborations	27	(34 personnel)
Co-Operative Awards	– SERC 3	(3 personnel)
	– MRC 2	(2 personnel)
SERC CASE Studentships	27	
Consultancies	45	
SERC Clubs	3	(4 in 1990)
LINK Schemes	2	(4 in 1990)

efforts of the company's Information and Library Service (ILS). This service is provided by four staff, who regularly provide a range of information to company employees, including:

– a daily journals list, which is a weekly accumulation of relevant journal articles;

– a monthly bulletin, including information about scientific papers ordered;

– a monthly patents bulletin, including information about patents ordered;

– a quarterly conferences bulletin, which includes details of conferences and those people attending from Celltech;

– a business information bulletin, which goes out every two months and includes details of the annual reports of interesting companies;

– a monthly new books acquired list – weekly press cuttings about Celltech.

The ILS keeps around 350 journal titles and has access to numerous online scientific and business databases; it also has a medical database on CD ROM. It keeps a number of databases related to particular projects. The CGRP project, for example, has a database of over 700 papers, and one other project has over 1000 papers stored. The project databases include information on patents, science and commercial information and academic liaisons. More of these databases are to be created in the future. Patent information is requested on average around 40 times, and around 40 to 50 online searches are conducted, each month. There are three or four requests for company profiles to be undertaken each month. The ILS keeps annual reports and market surveys on all its main competitors, and keeps full files on around another dozen companies in which it is interested. Information is also kept on the world top 40 biotechnology companies.

In addition to requests on patents from scientists, the company also acts proactively in its information search through its patent department. This department of four staff oversees the company's patenting interests, and has one member who has a permanent brief to watch the patenting of competing organisations. He spends a great deal of his time working in national patent offices. Scientific information is also brought into the company by the 50-odd seminars given each year by external speakers.

To the formal mechanisms for raising technological awareness discussed above must be added the informal ways in which scientists learn, through attending conferences and visiting other laboratories, and from the high number of academic visitors to Celltech.

Inter-Organisational Networking: the Case of the MRC

In addition to the strategy for inducing excellent internal communications and coherence, great emphasis has been placed on attempting to derive synergies between Celltech's own organisational structures and that of others, and ensuring that Celltech has employees receptive to, and capable of assimilating and developing external ideas. In this section, Celltech's relationship with the MRC, one of the most important to the company, will be examined. It will chart a lengthy, occasionally fraught, but essentially successful link which has depended on considerable adaptation by both organisations in both their formal systems and in the individual behaviour of scientists and managers. In many respects Celltech's close relationship with the MRC has provided the basis for its continuing strategy of high levels of collaboration, and assisted the development of the skills necessary to implement the strategy effectively.

The MRC played a central role in Celltech's formation. It had independently made the decision to support a British biotechnology company, so when the NEB approached the MRC with its plans in 1979, the MRC was very supportive of the project. The arrangement negotiated between the MRC and Celltech was a bold move by the MRC. Gerard Fairtlough describes the MRC's role in the formation of Celltech as 'imaginative and creative'. As we have seen it aroused opposition from the NRDC and the Treasury, and there are some claims that major UK pharmaceutical companies were highly antagonistic towards it. [8] It has to be seen as something of a gamble as Celltech was a risky enterprise, and had it gone wrong, it could have been harmful to the MRC. MRC participation gave the NEB plans enhanced credibility. The arrangement between the MRC and Celltech was one of the major factors attracting private sector investment into the company.

The MRC provided Celltech with its first products and continuing MRC/Celltech collaboration has given the company a number of other products and potential products. The company has recruited some of its leading scientists and managers from the MRC. The Secretary of the MRC, Sir James Gowans, sat on Celltech's Board from the company's inception until 1987 and continues his link with the company on a consultancy basis. Despite the high level of interaction between the two organisations, the relationship between the MRC and Celltech has not always been smooth.

On the same day that Celltech was formed, the company and the MRC signed a five-year Research and Marketing Agreement. Through the Agreement the MRC gave Celltech a first option to acquire exclusive world rights to develop and market all discoveries arising in its research laboratories in rDNA and cell fusion. Celltech undertook to ensure the speedy exploitation of such discoveries, to fund research projects in the MRC, and to feed back rewards from discoveries by means of the Celltech Fund, designed to fund Fellowships. Once the MRC had officially informed Celltech of an invention or a discovery, the company had six months to reach a decision on commercialisation. If the MRC was dissatisfied with Celltech's proposals for commercialisation it was free to approach another organisation.

Initially there was some disquiet amongst MRC staff about the Agreement. [9] A number of MRC administration staff recall opposition from some scientists within the MRC to the 'taint of commercialism'. There were also concerns about the ability of a small, inexperienced firm to maximise the returns from MRC inventions. These reservations by MRC staff were gradually overcome. This in part was due to the backing for the company given by senior MRC personnel, like Kirkman, Gowans, Brenner and Cesar Milstein, and the building up of trust by having MRC people so involved with the company. [10] Also important was Celltech's swift commercialisation of MRC discoveries like the anti-interferons and BGRs, and the symbolic significance of finance being fed back into the MRC by means of the Celltech Fund.

There was significant opposition to the Agreement from a number of sources outside the MRC. [11] Of particular importance was a recommendation by the House of Commons Select Committee on Education, Science and Arts, made in July 1982, which stated: 'We are opposed to exclusively (sic) in patent rights operated by any Research Council. We therefore recommend that the Government should urgently review the relationship between the MRC and Celltech, particularly as far as exclusivity in access to MRC funded research is concerned'. While there was concern at the breadth of the agreement and about Celltech's ability to deal effectively with discoveries in the increasingly important technologies of rDNA and MAbs, there was undoubtedly also a political knee-jerk reaction to the mention of 'monopoly'. Indeed, the Committee heard very favourable comments on Celltech's progress from the Royal Society, the Society for General Microbiology, BTG and the Secretary of State for Industry, Patrick Jenkin. Nevertheless, this recommendation put both Celltech and the MRC on the defensive. [12] The Government's eventual response to the recommendation, made in February 1983, was that as the Agreement was to be reviewed after three years, and exclusivity was one of the issues to be reviewed, and that there was no need to conclude the Agreement prematurely. [13]

For Celltech itself, relations with the MRC in the early years were mixed. To some extent the problems that emerged were inevitable. The expectations of two different organisations with differing ethos and aims had to converge. Practices and routines had to be established in an area where no previous role models existed. Generating good working arrangements between researchers required an understanding of the pressures facing others and this took time. Making time brought additional pressures. The decision that Celltech took to pursue rDNA research inevitably caused dissatisfaction amongst some MRC staff. [14] For those in the LMB, on the basis of whose research Celltech was initially formed, it appeared that the company was pursuing an avenue away from the LMB's distinctive advantage. Both Sydney Brenner and Cesar Milstein resigned from the Science Council in 1981. [15] One of the major criticisms levelled at the company in its first couple of years related to its inability to recruit a senior immunologist. This became a serious bone of contention within the company. Non-MRC scientists on the Science Council, and some Celltech staff, [16] echoed the MRC's concern. The company claimed that it was trying to recruit, but there was no-one of sufficient calibre available. It was not until Dr Lennox joined Celltech from the LMB in 1983 that this criticism was overcome.

The original 1980 agreement between Celltech and the MRC provided for a review of the arrangement after three years. This was duly conducted in 1983 and the exclusivity agreement was removed by mutual consent. Both parties claimed to have developed such close working relationships that a formal agreement was no longer necessary and a more general five-year agreement was signed. The company retained full rights to exploitation of ideas already adopted or MRC research currently being funded by Celltech. An immediate benefit to Celltech from this revamped agreement was a reduction in the amount payable to the NRDC from 1 per cent of turnover to 0.5 per cent (this lower rate was to be paid for five years). Another benefit to Celltech was that it removed the requirement to assess every discovery given to them, a time-consuming, and frequently wasted effort.

In practice, during this five-year agreement, there was a shift towards covering individual collaborations with specific collaboration agreements with their own provisions for funding, intellectual property rights, etc. The most notable developments in the Celltech-MRC interactions during the five years this Agreement ran were the collaborations (and occasional problems) in the field of antibody engineering. It is interesting to note that one particularly important collaboration for Celltech, with Dr Winter at the LMB on the 'humanising' of mouse MAbs, was initially troubled by many of the problems that had faced such agreements since the start: namely, a concern amongst scientists that their academic collaborations may adversely be affected by commercial links; and that Celltech may not be in the best position to offer the most generous financial rewards. Despite their experience in undertaking collaborations, and a convergence in expectations from both organisations about the form and benefits of collaborations, such linkages ultimately depend on good inter-personal relationships. These need to be re-established and nurtured each time collaborations take place between scientists for the first time.

A further five-year agreement between Celltech and the MRC, known as a 'Memorandum of Understanding', was signed in 1988. Essentially the main provision of the agreement was to ensure good communication between Celltech and the MRC such that the MRC will bring to Celltech's attention inventions which Celltech is well placed to commercialise, and Celltech will regularly update MRC of its technical (and, where relevant, business) focus. Communication is to be aided by regular quarterly meetings to review collaborations and to exchange information relevant to each party.

Misunderstandings and disagreements do still occasionally arise between the company and the MRC. One particular recent disagreement arose over Celltech's treatment of an arrangement with the MRC as a commercial contract rather than a collaborative research project. This episode highlights that even when two organisations are experienced at collaboration, when their aims are very different (commercial as opposed to academic), occasional misinterpretations will occur.

Celltech and the MRC are still enjoying highly successful joint efforts. At present Celltech is involved in a particularly important collaboration with the LMB, and there is considerable satisfaction expressed within the company at the smooth progress of this work.

Assessing the efficacy of the MRC/Celltech collaborative arrangements over the years is beyond the remit of this study. Nevertheless, many in Celltech and the MRC will argue that the linkages have been successful. The MRC provided the company with its first products and subsequently has contributed a substantial proportion of its science base. [17] Its present collaboration with the LMB is very important to the company. Opinion amongst those at the MRC during the period of the Celltech agreement unanimously endorse the value of the link with the company. The benefits described include: MRC scientists learning the value of industrial collaboration, and how best to go about it; increased stimulus to large companies to collaborate with the MRC; [18] and, the actual benefits accruing to individual scientists from the commercialisation of their discoveries. By 1986 it was estimated that the MRC was receiving £50,000 a year from Celltech in royalties; one-half of the Council's entire royalty income (Financial Times, 18 December 1986). In April 1988, the company claimed to have repaid the MRC nearly £900,000 in grants and royalties. [19] As a proportion of Celltech's total external R&D expenditure, the MRC is, however, declining in importance. Whereas for a number of years the MRC accounted for the vast majority of Celltech's external expenditure, this proportion is currently around one-sixth.

Given the extent of the benefits accrued by both parties, and bearing in mind the considerable efforts made by both to ensure the success of the collaboration, it is perhaps the case, as described by The Financial Times (12 November 1988) that the Celltech/MRC link has been 'one of the most successful transfers of academic science into industry'. These transfers have not been unproblematic. That they have occurred, and are continuing to occur, owes much to the perseverance of both organisations. Both have had to put in considerable efforts to ensure that the organisational systems are developed appropriate to the needs of collaborations, and that inter-personal expectations are converged so that good working relationships can be established. The relationship between Celltech and the MRC over the past decade highlights the many difficulties involved in collaboration; it also emphasises the value of long-term links and the need to put considerable effort into their effective management.

Footnotes

[1] See Pisano, Shan and Teece (1988), Chesnais (1988) and Hagedoorn and Schot (1988). It would appear, however, that these activities have been comparatively limited in the case of the major UK pharmaceutical companies.

[2] See Chapter 1.

[3] The majority of technological collaborations in biotechnology are between DBFs and large companies (see Chapter 1).

[4] The aim is to avoid grounds for complaint by sponsors of over emphasis on developing enabling technology to the detriment of project technology by means of tight contractual terms. There are also grounds for circumventing any complaints by the fact that although scientists are contracted to work a 37 hour week, many work over 50.

[5] Although the Science Council has little or no formal powers, its members have on occasion been extremely influential on decisions made at the top of the company. It has played an important role in informing the Board of the fast scientific progress the company has made, and the need for additional R&D funding. The Science Council actively promoted the management's plans for its new R&D building.

[6] External R&D expenditure is roughly one-fifth of internal R&D expenditure. Bearing in mind the fact that the majority of sponsored research is R rather than D, then external expenditure is roughly one-third of internal expenditure on the R side. The major Development project, in downstream processing, with the Scottish National Blood Transfusion Service involves just 5 per cent of external expenditure.

[7] Co-operative Awards are designed to encourage cooperation between academics and companies who wish to develop new products or processes requiring academic research. Provided the company puts in sufficient effort, material and expertise, the research councils will cover the bulk of direct costs. CASE Awards (Cooperative Awards in Science and Engineering) support research students who work on a project jointly devised and supervised by an academic department and a company. The company contributes a limited amount of the cost. The LINK scheme is a recent government programme devised to develop industrial collaboration in 'speculative' areas, such as biotechnology. SERC clubs, with which Celltech had an early

involvement (see Senker and Sharp, 1988), are joint industry/research council research promoting activities.

[8] The MRC has been criticised for 'locking-in' its expertise in Celltech, to the detriment of large companies (Senker and Sharp, 1988:58). It is interesting to compare the MRC's actions with the Science and Engineering Research Council's. Although the SERC had been involved in biotechnology since the late 1960s, and had organised an initiative on enzymes in the 1970s, it was not until 1981 that it established the Biotechnology Directorate as a means of encouraging technology transfer between academia and industry. In its early years the Directorate was dominated by large, established firms. In fact its first small firm member was Celltech which joined in 1984 (see Senker and Sharp, 1988). During the course of this study many senior people who were within the MRC at the time have recounted that the reason that it looked for a small firm to commercialise its biotechnology discoveries was that with limited exceptions large firms had shown no appetite for doing so, despite significant promptings. Some have gone on to argue that it was Celltech's formation that stimulated industry into dealing more with the MRC (see The Financial Times, 13 May 1983, p.25).

[9] New Scientist (13 November 1980) reported that at the MRC's press conference to launch its 1980 Annual Report some MRC staff reported 'that they were worried by two aspects of the relationship. First, there is no provision to plough back Celltech's profits into medical research, and secondly, scientists employed by the MRC might face a conflict of interest when setting the priorities of their own research'. In academic research, prestige, peer-recognition and promotion are often dependent on publishing findings before others. There were some fears that the Celltech arrangement might cause delay in publishing scientists' findings while the company sought to clarify its possible patenting position. Similarly, in biology and biochemistry, scientists frequently exchange materials and substances, and there was some concern that this exchange may be prohibited by the Celltech deal.

[10] The last two sat on Celltech's original Science Council. Milstein was renowned for his concern for 'pure' science. Brenner has been by many accounts very active in his support for the company.

[11] For example, Nature (26 February 1981) argued 'The agreement raises perturbing prospects of delays in the disclosure of scientific information'.

[12] See, for example, Gerard Fairtlough's letter to Nature, 12 August 1982.

[13] Observations by the Government on the 6th Report of the Committee in Session 1981-82, 15 February 1983.

[14] Sir Michael Stoker relates how, to his credit, Sir James Gowans exerted no pressure on the company to revert to research closer to the MRC's initial interests.

[15] The early loss of two such eminent scientists could potentially have produced adverse reaction to Celltech and the MRC Agreement. In the event, little comment was forthcoming.

[16] A number of present-day and previous Celltech employees have told me of the severe imbalance within the company between the young and inexperienced scientists working on MAbs and the strong rDNA team which included a number of world-famous scientists.

[17] In a presentation made to the British Association in September 1988, Gerard Fairtlough said 'it is difficult for me to imagine how Celltech would have reached its

present position of leadership in mammalian cell culture for the production of complex proteins and in the genetic engineering of monoclonal antibodies unless we had been fairly close to the MRC Laboratory of Molecular Biology'.

[18] Sir James Gowans was quoted in 1983 as saying it was only the MRC/Celltech agreement which awakened industry to the commercial possibilities of MAbs (The Financial Times, 23 May 1983). In October 1986 the MRC established its Collaborative Centre to liaise with industry.

[19] Gerard Fairtlough, speech celebrating the opening of Celltech's new R&D building, 15 April 1988.

Chapter 9
The Strategic Management of Learning

Introduction

There is a large and growing literature on technology strategies in firms (Porter, 1985; Horwitch, 1986; Teece, 1987; Dodgson, 1989; Loveridge and Pitt, 1990). While the concept of technological learning is not centrally addressed by this literature, learning is often implicitly assumed to be one of the components of strategy, and the ability of firms to learn about and adapt to new technologies is identified as a key feature affecting firms' competitiveness. Technological learning is addressed by a range of academic literatures, such as innovation studies, industrial economics, organisational behaviour, and management studies, but it rarely enjoys systematic and coherent consideration. The study of Celltech contributes to the understanding of the processes and intent of technological learning, and its relationship with technology strategy.

Dedicated biotechnology firms' past and future growth rests with strategies for fast technological learning: the ability to acquire, develop and operationalise technological know-how quicker than their competitors. It is this which potentially gives DBFs a comparative advantage over large multinational companies, and has been a major stimulus to high levels of interaction between them and large established firms. The ability of DBFs to establish creative, well organised and highly motivated R&D departments, to work effectively with the science base, and to benefit from collaboration with larger industrial partners are all features of technological learning, and are crucial to DBFs' competitiveness.

Strategy: Conceptual Background

Figure 23 conceptualises the intermediary role that strategy plays in linking a company with its competitive and technological environment in the specific circumstances of a science-based company. In this model (based on Chandler (1962) and Snow and Miles (1983)) strategy is an analytical tool which can be considered as an interjacent filter between the complex interrelationships of a company's environment and its internal activities, organisation and capabilities. Thus strategy is a means by which the links between an organisation's strengths and weaknesses and its environmental opportunities and threats (competitive, scientific, regulatory etc.), can be conceptualised.

In a normative sense, the model describes how, in order to be effective, strategy should be aligned to its competitive and scientific environment and arranged in line with its competences. The relationship is multi-directional: strategy generally responds to the environment, but also can be formulated in order to alter it. Similarly, the relationship between the company's strategy and activities is two-way. Although a company's strategy influences the activities and organisation of the firm (following Chandler's dictum that structure follows strategy), the capacities of the firm (particularly its knowledge and skills bases (see Georgiou et al, 1986)) affects the strategies likely to be adopted. The model also emphasises that for strategy to be reflexive and responsive, it is essential that there are suitable information flows into and about the various activities of the company, and that these are communicated fully throughout the organisation.

When firms are operating in rapidly changing circumstances, and when being first is often the key to competitive advantage, as is the case for DBFs with the race to achieve protectable patent positions, then there is a premium placed on firms' abilities rapidly to comprehend the changes taking place, and to utilise this knowledge within the organisation. Awareness of strategic opportunities has, of course, to be translated into an organisational response. This is facilitated by organisational and interfunctional cohesiveness and by good information flows and communication within the firm. In essence considerable value can accrue to the firm most capable of aligning its strategy with its environment and arranging its activities in line with its strategy.

Figure 23

Conceptualising Strategy

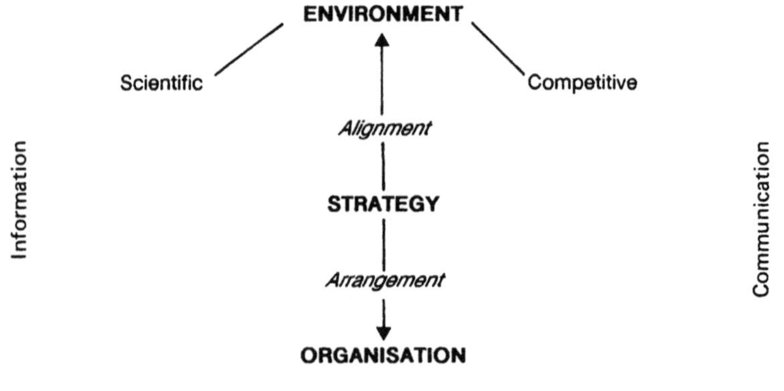

Learning: A Brief Review of the Literature

A number of examples illustrate the ubiquity of concepts of learning. Learning is central to Cyert and March's theory of the firm. "A theory of long-term behaviour in organisations must contain a theory of how organisations learn, unlearn, and re-learn" (Cyert and March, 1963). The differing ability to stimulate and accumulate technological knowledge in changing technological and competitive circumstances is a feature distinguishing Burns and Stalker's classic (1962) analysis of organic and mechanistic organisational forms. Organisational learning is the focus of Argyris and Shon (1978), and is a feature of Ansoff's (1982) contribution from a strategic management perspective. Concepts such as the 'learning curve' are widely used in managerial literature, and its use as an analytical tool is extensive, actively promoted by consultancy companies such as the Boston Consulting Group. Learning features in some managerial literature on new product development (Maidique and Zirger, 1985). Recently economic theory's consideration of technological and industrial development has begun to emphasise the importance of learning. "...the features of the evolution of each industry are...'ordered' by the patterns of learning, and by the ways the latter influence the competitive process" (Dosi, 1988).

Numerous stimuli to learning in firms have been suggested, including both environmental and intrinsic factors. Freeman and Perez (1988) ascribe renewed learning activity in firms – about organisational forms and the skills required – to the environmental turbulence engendered by changes in 'techno-economic paradigm'. Triggers to learning within firms include existing problems (Hedberg, 1981); affluence (Cyert and March, 1963); and executive succession (Tushman, Virany and Romanelli, 1986).

The literature on technological learning has been marked by considerable heterogeneity, often compounded by definitional differences. It has tended to focus on learning in production and post-production, rather than on R&D, and rather narrowly has concentrated on the learning that occurs as a 'natural' result of cumulated experience. To some extent this emphasis on manufacturing learning follows the impact of the work of Arrow (1962), and his contention that firms develop increasing skills in production with the result of reducing real labour costs per unit of production ('learning-by doing'). While this thesis has its shortcomings, it has attracted considerable attention, particularly amongst economists. Another feature of the literature on learning is that it primarily focuses on established rather than newer firms.

Despite its diversity and shortcomings, some of the literature on learning, including that focusing on learning in the narrow sense of learning-by-doing, provides a useful guide for the broader definition of learning used here. It is intended to discuss technological learning in both R&D and production. A wide range of factors which contribute to learning are examined, ranging from learning through experience, on the one hand, to recruitment and training policies to augment learning, on the other.

A number of features of the literature on learning, and its criticisms, inform the approach taken here. Bell and Scott-Kemmis's (1990) criticisms of the causal factors ascribed by the 'learning-by-doing' thesis, are helpful. They argue that 'learning-by-doing' has become accepted as a truism amongst some economists, and that the cause of this form of learning has not been examined in any depth. In their view this approach is fundamentally flawed as it assumes that learning is automatic and costless and it ignores the continuance of technical change. They contend that incremental productivity is in fact determined by incremental technological change, and requires active managerial inputs. This book adopts a position contrary to that held by many industrial economists in arguing that learning occurs through purposive activity, and does not occur automatically.

Rosenberg (1982) focuses his analysis mainly on learning-by-doing and learning-by-using (learning from the use of a product), but he also addresses learning in Research and Development (R&D).

"R&D is a learning process in the generation of new technologies. At the basic end of the spectrum, the learning process involves the acquisition of knowledge concerning the laws of nature. Some of this knowledge turns out to have useful applications to productive activity. At the development end of R&D is a learning process that consists of searching out and discovering the optimal design characteristics of a product. At this stage, the learning is oriented toward the commercial dimensions of the innovation process: discovering the nature and combination of product characteristics desired in the market ... and incorporating these in a final product in ways that take account scientific and engineering knowledge" (Rosenberg, 1982).

Thus learning is not uniform in purpose or nature, even within functions. Learning has a 'strategic' element, characterised by search routines designed to bring knowledge into the firm, or to generate it internally, about a scientific field or a technology. It also has a 'tactical' element, which essentially is concerned with the provision of a particular product or service to a customer. Learning is at once concerned with science and commerce. But the firm has an important interface function as Rosenberg stresses: using the scientific/engineering knowledge to produce products society wants.

With reference to the literature cited above, and for the purposes of this book, learning is defined as the ways firms build and supplement their knowledge-bases about technologies, products and processes, and develop and improve the use of the broad skills of their workforces.

Conceptualising Learning

Learning is a heterogenous concept, it has differing aims, focus and involves different processes.

– The Aims of Learning:

As Harris and Mowery (1990) tell us:

"When firms generate innovation, or when they adopt innovations generated by others (and whether they act deliberately of 'accidently') there is, at minimum, the opportunity for learning. In other words there are two separable outcomes of innovation processes, the innovation itself and the learning about the innovation and innovation process".

In this way the aims of learning may be strategic or tactical. Tactical learning may be described as that which has an immediate problem-solving nature. It may relate, for example, to a product or to an operational problem. The learning required may involve recourse to an R&D or Engineering department, or to an external source, perhaps through the use of a consultant, research association, university or even collaboration between firms. The aim of the learning is identifiable, and the time-scale is short and prescribed. Any improvement in the innovative capabilities in the firm are project specific. Strategic learning extends beyond immediate issues and involves firms developing skills and competences which provide the basis for future, perhaps unforeseen, projects. Strategic learning attempts to maximise the returns from the opportunities that the innovation process provides.

In the sense used here, strategic and tactical learning are not synonymous with research as opposed to development, as both these activities may be strategic or tactical in focus. Strategic and tactical learning are more closely analogous with what Patel and Pavitt (1988) refer to as myopic and dynamic firms. The former category is concerned only with the immediate problems, the latter takes the broader, long-term view of technological accumulation: that is, technology is acquired for its potentiality. Argyris and Shon's (1978) analytical distinction between single-loop and double-loop learning and deutero-learning is also useful. In a very simplified way, single-loop learning involves feedback into current decision-making practices so as to improve future decisions. Double-loop learning involves organisations questioning the whole basis of their decision-making practices and involves wholesale modification within the company. Deutero-learning examines the whole process of learning within the company, and can be viewed as 'learning about learning'. Tactical learning is equivalent to single-loop learning. Strategic learning includes both double-loop and deutero-learning.

– The Focuses of Learning:

Learning does, of course, have different focuses. These may include R&D, production, and managerial/organisational concerns, but this list is nowhere near exclusive. Learning is important throughout the activities of a firm, and the focuses of learning change over time. As Hamilton, Vila and Dibner's (1990) study of a sample of 87 US DBFs shows, these firms' activities steadily shifted from science to commerce over time. The focuses of learning adapt accordingly.

The strategies adopted by DBFs, rapidly altering their learning focuses, reflect their attempt to respond more quickly than large firms to changing technological and competitive circumstances. Thus Celltech's and other DBFs' original strategies required priority be placed on building R&D skills. Subsequently emphasis was placed on manufacturing competences. Currently these companies are adding to these skills a growing ability in medical, regulatory affairs and marketing functions.

Two particularly important focuses of learning are understanding environmental change and developing managers' and scientists' competences. A central focus of strategic learning is gaining an understanding of the opportunities that changes in the firm's competitive and technological environment provides. In the case of DBFs, and in all high technology companies, this requires the detection of technological shifts. One of the major obstacles to successful growth in high technology firms, as described by Hambrick and Crozier (1985), is the recruitment of people who, with continuing training, will grow into their new and increasing responsibilities as the company grows.

– Learning Processes:

Technological learning differs broadly in its nature. While a number of typologies have been suggested on the nature of learning (see Valentin, 1989), three broad kinds of learning are distinguished here: search, accumulation and diffusion, and review.

Search: Search involves seeking knowledge, either internally through the firm's own activities, for example, in R&D or engineering, or externally, for example, from scientific and technological scanning. Internal search may be directed into discrete technologies or projects, and may also include efforts to find linkages between different scientific and technological activities. Search activities may be strategic or tactical; they may be concerned with a particular problem to be solved or may be directed towards the long term acquisition of potentially useful knowledge.

Accumulation and Diffusion: Once information has been sought and found it then needs to be disseminated into the firm's stock of knowledge. Accumulation of know-how again has strategic and tactical elements. Learning about a tactical problem should be accumulated and diffused within the firm so as to prevent repetition of search activities. Learning about strategic issues should be accumulated and diffused to inform future strategic plans and departures in the various parts of the firm.

Review: An important aspect of learning is reviewing what you know. Such reviews can assist strategic decision-making, and can highlight deficiencies in the search and accumulation and diffusion process. An aspect of review learning is learning from mistakes or failures.

Learning in Celltech

Learning as a feature of strategy is driven by competitive forces and in Celltech's case by its attempt to achieve a competitive position in an emerging technology through its organisational specialisms. The company's corporate strategy has been adapted over the ten years of its existence, and mirrors the way that leading US DBFs have developed. Its mission statements have reflected the way that the technological, competitive and regulatory circumstances have altered during this period. Essentially the company's strategy has changed from being a research company to a research and manufacturing company to a fully integrated biopharmaceuticals company developing, manufacturing and marketing drugs. Its comparative advantage over large pharmaceutical firms lay firstly in its linkages with the science-base, in particular with the UK's Medical Research Council (MRC). As it began to develop its internal skills, it identified the scientific feasibility and commercial potential of the manufacture of biotechnology products. It rapidly developed a world-leading capacity in its manufacturing facilities ahead of its major competitors. More recently it is attempting to fully integrate its activities on the basis of its R&D base, its manufacturing know-how, and in the belief that it can identify market niches which it can fill quicker and more cheaply than its large firm competitors.

A number of measures – apart from the company's rapid growth – have been suggested as indicators of Celltech's ability to learn about, develop and utilise technology quickly. For example: the registration of its first patent, based on its own research, six months after its formation; the indicator that it currently has more biotechnology patent applications pending than the UK's Wellcome and Glaxo, and almost as many as Beecham; and its rapid development of a world-leading capacity for the bulk manufacture of biotechnology products.

Celltech's learning efforts have to be placed in the context of the legal/regulatory system in biotechnology. There are many unresolved regulatory issues and questions concerning intellectual property rights in biotechnology. Nevertheless, great emphasis is placed on the attainment of strong patent positions. There is considerable advantage to be achieved for firms to be first to register and protect their discoveries. In this respect the ability to learn quickly is at a premium as the firm that can translate science into products or processes first has the opportunity to establish a patent protected position.

The relationship between technological learning and adaptive strategies is not peculiar to Celltech, and is identified by Loveridge and Pitt (1990) as being applicable to other, larger firms:

"..effective strategies of technological innovation benefit from, indeed may necessitate an adaptive, learning orientation in which strategists and managers would do well to come to terms with the reality that they are in full control of the firm's destiny. Thus adaptive, opportunistic knowledge appropriation as a means of effective learning is the key to technological transformation".

Celltech provides an excellent example of a company for whom the strategy of technological learning has allowed an adaptive commercial strategy to be pursued. Underpinning the strategy has been a concern to develop the skills of the company's management. Following sections will describe and analyse how Celltech's managers have identified strategic aims, progressively increased the focuses of learning, and used learning processes.

Management Structures for Strategic Decision-Making

In a study of 38 leading European high technology small and medium-sized firms (Dodgson and Rothwell, 1989b) it was found that a key factor underlying their success was the capacity for their management to plan strategically. These companies had clear visions of what markets were to be addressed, and what was needed, in terms of R&D, personnel, marketing and other resources, in order to reach these markets with winning products. The absence of these strategic management skills are one of the major barriers to growth in small firms (ACOST, 1990). Celltech provides an example of a smaller firm whose management dedicates considerable time and effort to consideration of strategic issues. This has had a number of benefits. For example, the company's selection of technologies in R&D has proven highly effective. As figure 15 shows, the technologies originally selected have invariably been found to have long-term significance. [1]

There are a number of strategic and operational committees directing the company's activities. [2] The primary decision-making body, the Board of Directors, includes six Celltech executives in its 12 members. Strategic decision-making is assisted by the company's Science Council (see Chapter 8), and by a Strategy Committee. The Strategy Committee meets regularly to analyse and make decisions on key strategic issues. As an important input into the strategic decision-making process the company organises so-called 'Management Away-Days'. [3] The main purpose of Away-Days is to give managers time off from immediate pressures to enable them to consider strategic issues.

The company has, since its inception, produced an Annual Strategic Plan. This five-year planning procedure is coordinated by the Company Secretary. The process of producing the plan used to start early in each new year, and drafts were commented upon by the Strategy Committee until the document was presented for approval by the July Board meeting. Recently the procedure has changed so that it involves a continuous process of input and feedback throughout the year from several standing sub-groups examining particularly important issues.

Operations within Celltech are managed by a Company Operations Group, chaired by the Chief Operating Officer. Day-to-day co-ordination of R&D activities is undertaken by an R&D Operations Group, chaired by the Director of Development. New projects are scrutinised by a Project Planning Group, run by a Project Planning Manager, and including the Directors of Research and Development and a number of coopted members, including managers from Sales and Marketing. Its role is to check, comment upon and improve project plans while taking account of the balance between the company's own-risk and contract projects. It is designed largely to ensure that there is a single, consistent mechanism for approving and prioritising all the company's project activities in line with company objectives. It reports to the Company Operations Group.

The level of formality of these organisational structures may appear unusual in a medium-sized company. However, consideration has to be made of the purpose of these structures, which are not only, and in some cases not primarily, decision-making bodies, but important mechanisms for information sharing and consensus forming. The company's 'paper intensiveness' – minutes, agendas, discussion papers, and plans tend to adorn managers' and scientists' desks – also is an important mechanism of information sharing. One of the strengths of having good and broad communications paths between the 'strategic' and 'operational' elements of the company (managers often have responsibilities in both areas), is that both are informed by the decisions and actions of the other.

Strategic management has, of course, to be complemented by effective day-to-day operational management. In this respect Celltech has, according to some in the company, occasionally not always been best served. Nevertheless, considerable changes have taken place within the company as it has grown in order to ensure that this complementarity is achieved. As the company has evolved executive operational power has devolved. From being entrepreneurially controlled in its operations in the first years of its existence, it is now run by a broader franchise of decision-makers. Similarly, changes have taken place over time in the relative importance of various management functions and concomitant influence of particular managers. First priority lay in building up a research function, [4] and we have seen that the relative emphasis placed on different fields of activity within this function has changed over time. Emphasis then changed to building up marketing expertise. Development and manufacturing have been prioritised at various times. Currently the emphasis lies in building up the broad capacity for pharmaceutical development and marketing.

All these changes have had consequences for the power and authority of individuals. It is perhaps testament to the company's management style that these changes have not been traumatic, and with few exceptions have not involved resignations of people who have a great deal to offer the company (thereby ensuring, of course, continuity amongst employees skilled in the company's specialisms). These transitions and changes in emphasis are inevitable in growing companies, and can act as a constraint to growth if they are not overcome (Hambrick and Crozier, 1985). Perhaps the most important transition that needs to take place is at the most senior level, with the necessary change from 'entrepreneurial' management provided by a company founder, to the 'professional' management provided by functional specialists. [5] Celltech achieved this difficult and crucial transition in 1985/86, most obviously when David Gration assumed the role of chief operating officer, and he reorganised the company on its current functional basis. [6]

Management Development

Crucial to Celltech's continuing growth has been the development of managerial skills within the company. In its first few years the company had only one management representative amongst its eight Board Directors; in 1990 it had six managers on its twelve-man Board. This shows how the company is increasingly being directed by managers whose talents have emerged while working for it. It is the strategy of selective recruitment and development of people with the ability to take on novel and extensive responsibilities that enables firms to overcome the 'critical thresholds' or 'transition periods' which affect growing firms (Metcalfe and Tiler, 1989; Scott and Bruce, 1987; Dodgson and Rothwell, 1989b).

It has been the policy of Celltech's leading managers from its inception to enfranchise as many suitably qualified and experienced people into the decision-making process as possible. [7] This has involved the embodiment of management skills where applicable in scientists. Many scientists have responsibility for budgetary controls, and emphasis is placed on training them in the broad range of skills beyond research and project management.

An Example of Strategic Management: The Development of the Pharmaceutical Strategy

Perhaps the most pronounced example of strategic leadership that Celltech has displayed is its decision to pursue its pharmaceutical strategy. The extent to which the company was forward-looking in 1985, when the strategy was first formally suggested, is revealed by the fact that its 1985 strategy document made it clear that the company could successfully pursue its existing strategy for the planning period of five years. However, the document noted that in the 1990s large pharmaceutical firms were likely to expand their own biotechnology activities in R&D and manufacturing, and this would adversely affect Celltech's position as a contract research/manufacturing company, and its ability to retain key staff. Particularly of concern was the way Japanese companies were targeting biotechnology as a way of achieving strong positions in the world pharmaceutical industry.

The 1985 document describes a strategy which reflected both environmental and internal issues. It raised a number of features of biotechnology-based drugs for consideration. These included: the extent to which discoveries could be protected by patenting or by keeping know-how in-house (this was, however, left as something of an open question); the problem of drug delivery (biopharmaceuticals generally have to be injected), where Celltech possessed a potential advantage; the selection of drugs to be targeted, and decisions on how they could be marketed; the question of whether a small company could compete with giant multinationals; and whether Celltech could attract the requisite skills. On the last two questions, the conclusions were that to adopt a strategy aimed at becoming a manufacturer of pharmaceuticals required new specialists in medicine, pharmacy, pharmacology, and international distribution, and that they could, on the basis of Celltech's existing successes in this field, be recruited, and that management could ensure they would be highly motivated. It was also argued that small firms could compete with large established firms, as the electronics industry had shown, but 'the newcomers must behave in a different way from the established firms, avoiding some of their overheads, their bureaucratic structures and their internal politics'. Furthermore, it argued that large firms had continued to use their traditional methods of drug discovery, without realising the full potential of the advances in molecular and cell biology. Also important in this respect was the way large firms were acquiring new biotechnology firms (e.g. Hybritech, Genetic Systems and Biogen), which was believed to indicate that they found it harder to adapt existing organisations than to acquire external competences.

On the question of therapeutic areas to be targeted, it was decided that early focus would be very beneficial, and that they should be selected according to market significance and in physiological areas where Celltech had particular strength, i.e. in molecular biology, molecular genetics, immunology or cell culture. Opinions were sought on targets from a variety of sources, and a market research exercise was undertaken. A number of potential products were identified.

Two broad courses of entry into pharmaceuticals were considered: going it alone in a number of years, or forming a joint venture to do so within a shorter time period. Going it alone required the reservation of a portfolio of potential products which would not be licensed to others at least within a core geographical area. When a suitable portfolio had been built up, and when the company had a sufficient track record to raise substantial further capital, then Celltech could enter pharmaceuticals by itself. This was believed to be the less risky option, as the reserved portfolio provided the option of a fall-back to licensing income. The second course considered a number of joint venture structures, and a number of UK and Japanese partners. The document recommended to the Board that both options be considered, and while the 'reserved portfolio' plan be pursued in the immediate future, exploratory discussions should be held with a limited number of potential partners.

The strategic planning document therefore addressed all the broad questions necessary in an exercise of this nature. It asked what future changes are likely to affect the company's competitiveness, what needed to be done to maintain and improve competitiveness, and how these changes were to be brought about.

Conceptualising the Managerial Task of Technological Learning

The effective management of technological development is, of course, central to the competitiveness of DBFs. As Robert Fildes of Cetus puts it:

"The first challenge is managing technological excellence. A company in this field, and especially a small company, really only has one ace in the hole competitively speaking. That ace is its technical prowess.." Fildes (1990)

Figure 24 offers a conceptualisation of some of the major management learning processes and tasks concerning technological development and use. Some of these have already been discussed. A particularly important part of strategic learning is gaining knowledge of the opportunities that changes in the competitive and technological environment provide. This requires the detection of technological shifts which threaten existing ways of doing things and provide the potential for alternatives. As we have seen in Chapter 1, DBFs came into being on the basis of a technological shift, and management are highly informed in this respect. Knowledge in selecting projects has also been discussed, and we have seen Celltech's success in selecting technologies in R&D and initial failures in selecting projects for commercialisation. The company learned from these failures and has become more skilled at project selection. The ability to learn from R&D and from manufacturing has been discussed in Chapters 7 and 8. The strategic element of this learning is particularly clearly revealed in the company's attempts to maximise its 'enabling' technology on the basis of its contract research. Human resource management and organisational structure issues have been considered in Chapter 6.

A number of examples of the way that Celltech undergoes its search, accumulation and diffusion, and review processes are provided below.

Examples of Celltech's search activities include the following:

– The Science Council of foremost British scientists which includes amongst its functions bringing new ideas into Celltech and filtering new ideas from any source.

– Encouraging scientists to retain strong links with the science-base. Celltech scientists spend a considerable amount of time visiting laboratories and attending conferences worldwide. Many academics from around the world visit Celltech, and we have seen the high number of seminars given by visiting scientists (around 50 a year).

– The company's highly proactive Information and Library Service (ILS) undertakes extensive searches.

– The patent department is highly active in its search activities.

Examples of Celltech's accumulation and diffusion activities include the following:

– The very high number of intra- and inter-functional meetings to assist communication. We have seen how 50 seminars are held each year by Celltech staff to assist internal information flow.

– The provision by the ILS of frequent reports to staff on: all patents, articles and books requested by people in the company; details of conferences attended; and details of company information requested.

– The compilation by the ILS of databases of all information requested on a particular project.

Examples of Celltech's review processes include the following:

Figure 24

Conceptualising the Management of Technological Learning

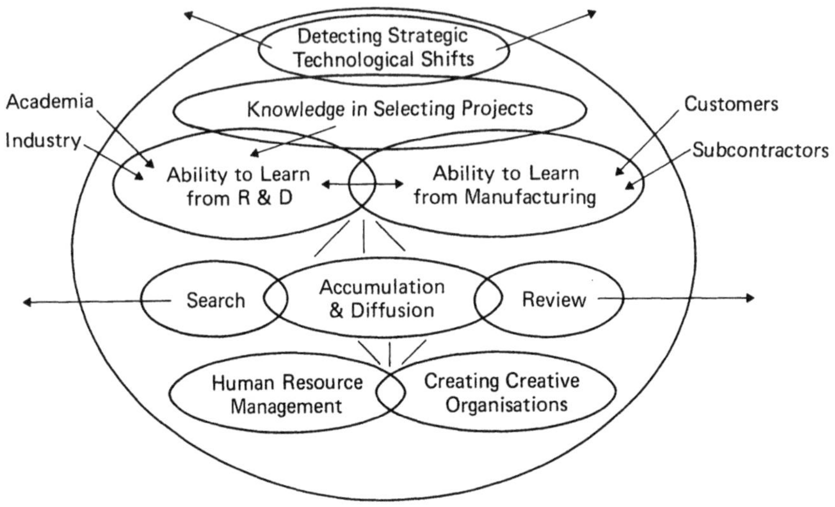

– Continual year-round strategic management exercises designed to inform annual strategic plans, and involving considerable emphasis on assessment of current strengths and weaknesses.

– Exercises to examine past failures in order to prevent their repetition.

– The Science Council undertakes reviews of major topics (such as large projects) at the request of the company's R&D management.

Learning by failing is an important aspect of any firm's development. Systematic evaluation of past mistakes is a useful mechanism for avoiding their future repetition. We have seen in the case of chymosin that Celltech has occasionally assessed where mistakes have been made. This procedure has been undertaken for a number of projects which did not turn out as well as expected. The company has also in the past undertaken exercises which examined other sorts of failure, attempting to improve, for example, presentations made to potential collaborators. In Chapter 8 one particular example of a continuing process of learning was examined: the relationship between Celltech and the MRC. This highlights the difficulties in establishing working relationships between organisations with very different structures and aims. It also emphasises the importance of perseverance, and continually learning from mistakes.

One other element of the management task of learning warrants consideration: the balance between internal and external learning. Celltech has depended on both internal and external learning, and although both are important the balance between them varies over time. The company reveals how a crucial aspect of the successful management of strategy and learning is the way that complementarities between external and internal learning is achieved. Decisions have to be made on the sources of learning to avoid repetition, and to maximise the comparative advantages of research in-house and externally. The latter decision requires consideration not only of where technological advantage lies, but also of how future outcomes of research are to be appropriated, ie who owns the intellectual property rights. In this respect, technological learning is again shown to be a central strategic management consideration.

Conclusions

Technological learning is a multi-faceted concept, and this chapter has offered a distinction between the aims, focus and processes of learning, and a conceptualisation of its management. In conditions of competitive and technological turbulence corporate strategies may increase the emphasis on those elements dedicated towards learning in all its forms: tactical and strategic; search, accumulation and diffusion, and review. Such strategies are purposive, have internal and external elements, and are closely allied to strategies concerned with human resource management and organisational structures.

A major feature of Celltech's technology strategy is its emphasis on, and encouragement of, learning. Priority has been placed on building the company's distinctive core technological competences: originally in its R&D skills in rDNA and MAbs; and subsequently its bulk production of proteins through cell culture and antibody engineering. This has involved considerable internal learning, for example, in the manufacturing function; and considerable external learning, for example, in antibody engineering. Emphasis has also been placed on extending these core activities, and attempting to learn about new opportunities. The company has shown from the start that it is adept at responding quickly. Learning from successes and failures is essential in fast changing environments. The science and technology of biotechnology, and its potential markets, have altered dramatically in the short period of Celltech's existence. The ability of the company to adjust to these changes, and indeed to benefit from them emphasises the importance of adaptability and the ability to learn.

Effective technology strategies often require a high level of confluence between the technological elements involved and human resource development strategies (Senker and Brady, 1989). The organisational structures of firms inevitably affect their efficiency in, for example, their ability to communicate and attain synergies. This applies to the organisational structure of a firm as a whole and to its constituent groupings. As Orsenigo (1989) puts it:

"...a fundamental property of learning processes is that they are intrinsically social and collective phenomena. Individuals rarely act in isolation, but interact with other people.... Moreover, given that individual agents are characterised by limited capabilities, the solution of complex problems requires in most cases the coordination and integration of different skills and heterogenous pieces of knowledge in a coherent framework" The extent of the interdependency of technology strategy and learning, organisational structure and human resource management is clearly shown in Celltech's case.

Footnotes

[1] It could be argued that the emphasis on strategic management detracted in the early years of the company from more tactical decisions, such as project selection. As we have seen it was less successful in selecting projects in the first few years.

[2] The Executive Directors are: the chief executive, the chief operating officer, and the Directors of Scientific Affairs, Finance, Development, and Sales and Marketing.

The Strategy Committee comprises the six Executive Directors, the Directors of Research, Academic Liaison, Manufacturing and the Medical Director, the Company Secretary and a senior scientist.

The Company Operations Group includes the chief operating officer, the Directors of Scientific Affairs, Manufacturing, Development, Research, Finance, Sales and Marketing, the Medical Director and the Company Secretary.
The R&D Operations Group includes the Directors of Development, Research, Academic Liaison, and the Medical Director, the Head of Research Project Management and a senior Project Manager, and the Heads of R&D functional sub-divisions.

[3] Management Away-Days are quarterly meetings of the Strategy Committee and involve an overnight stay in a hotel. On the first evening a presentation is made by an expert in a particular field (a City analyst, for example) and reviews of particular important issues are undertaken. The following day a number of selected topics are considered. Small groups are syndicated to consider various aspects of the topics, then all the groups meet together to discuss their ideas. In addition to the members of the Strategy Committee, three or four other specially invited Celltech staff attend Away-Days.

[4] Its first three employees – Fairtlough, Carey and Vaughan – were all biochemists, and its first recruits were scientists.

[5] In the study of a sample of leading European high technology small and medium-sized firms it was found that a major reason for their success was their ability to make the transition from entrepreneurial to professional management (Dodgson and Rothwell, 1989b). It would be a mistake to describe Gerard Fairtlough as a conventional entrepreneur. Although entrepreneurial, his motivation derives from the vision he has for Celltech as a new form of organisation and for British biotechnology in general (motives which apparently confuse some people). He is also, of course, a highly experienced professional manager.

[6] Gerard Fairtlough relates how this period of 'letting go' of operational control was personally difficult for him, despite his realisation of the necessity for it. No other person interviewed mentioned this transition, so in itself it obviously caused few organisational or operational difficulties, which reveals the professional manner in which it was undertaken.

[7] Celltech's chief executive, Gerard Fairtlough, has published extensively on organisational innovation, and his views have been considered in Chapter 6. By way of example, in one of his papers he emphasises the importance for organisational innovation of: encouraging articulation amongst people, and the open analysis of problems; learning from others; and gaining broad consensus about new ideas (Fairtlough, 1986). His intellectual interest in the subject has influenced Celltech's structure and style.

Chapter 10
Lessons From the Learning Firm

Introduction

In the previous chapters an account has been provided of the development of an exemplar DBF. Considerable detail has been provided on the way the company was formed and developed, with particular attention being placed on the way that technology was acquired, developed and utilised. The way the company learns has been argued to be an important facet of its competitiveness and a focus of its strategy. This chapter will, through reference to Celltech, examine the major factors to have affected the development of DBFs and the role that they have played in the industrialisation of biotechnology. It will also refer to a number of the problems encountered by small, fast growing, high technology companies. It was argued in Chapter 2 that Celltech's successes and failures have to be placed in the British context, and reference will be made to the national influences which have specifically affected the company.

Successes and Failures

Biotechnology has in many ways grown in the shadow of the information technology industry. Expectations were of high growth, and of a number of new start-up firms becoming major industrial players, such as DEC and Apple have managed to become in IT. Yet as Calori (1990) tells us "Success stories of small pioneers who became independent major competitors at the emerging phase (of new industries) are extremely rare". However, DBFs have played an important, if not the major, role in the development of biotechnology to date.

They have formed strong links with academia, taking science and recruiting scientists from universities and teaching hospitals, and directing efforts towards the creation of biotechnology products and services. Apart from undertaking considerable R&D themselves, they have also worked extensively with large, established firms in a variety of R&D contracts, joint R&D projects, and joint ventures. Some of the leading DBFs have strategies for becoming fully integrated companies, developing, manufacturing and marketing biopharmaceuticals. A few leading companies have actual therapeutic products on the market.

However, in the light of the recent acquisition of Genentech, it appears that very few of these firms will have the opportunity to become independent pharmaceutical firms.

Throughout this book it has been argued that Celltech is an exemplar DBF. To illustrate the factors influencing DBFs' patterns of growth Celltech's contributions and achievements will now be discussed, and those factors which have contributed to success and have constrained development will be highlighted.

If Celltech is judged according to the narrow criterion of profitable returns for its shareholders, then it, like the vast majority of DBFs, has not been a success. It has only created limited profits, and has little chance of returning substantial profits for some years. The high levels of R&D expenditure necessary to fund drug development have inevitably suppressed and delayed profitability. If growth in market capitalisation is considered, then the company's value has increased ten-fold in ten years. [1] If broader criteria are selected then Celltech's successes are easily identified. Celltech is Europe's leading independent biotechnology company and is amongst the world's top dozen such companies. With the opportunities in Europe provided by 1992, and the enormous potential of biotechnology, this position provides the company with a considerable advantage. The company has successfully commercialised science and technology from academia. Exports account for up to 90 per cent of the company's steadily increasing sales. It has marshalled a considerable, and highly professional, research and manufacturing resources, both in human skills and plant and equipment, and has created some world-leading technology. This resource has proved attractive to many of the world's largest pharmaceutical companies. Major US, Japanese and European drug companies have undertaken business with Celltech because it possesses skills and capabilities they do not.

Celltech's achievements have also to be more widely assessed. The company has undoubtedly improved the opportunities for technology transfer from academia to industry by providing an additional conduit for those academics wishing to commercialise their discoveries. The formation of the company played an important role in changing the previous rigid system for commercialising UK academic science. Its creation effectively broke the NRDC's monopoly on patenting decisions about scientific discoveries funded by the government. In many ways the company's formation was instrumental in altering the technology transfer mechanisms from academia to industry. The MRC, the company's partner in creating this departure, has particularly benefited from its existence. According to many in the MRC, Celltech has stimulated academics to place greater consideration on the commercial implications of their work. The MRC's Collaborative Centre provides the most tangible evidence of a major transition in attitudes by a Research Council to industry. As we have seen, however, Celltech's academic linkages have been extensive and not been limited to the MRC.

In a recent article in the Financial Times, on the British scene Celltech has been described as the 'grandfather of the biotechnology business'. In this role it has provided an opportunity for those scientists wishing to use and develop their skills in Britain. [2] Furthermore, the company has also provided a 'training school' for scientists, some of whom have left the company and are using their skills elsewhere: for example, in large firms developing biotechnology interests; in spin-off firms; or in technology transfer organisations. [3]

For Celltech, and for other DBFs, it is inappropriate to adopt a narrow assessment of their achievements. Although few have made profits, they have played a centrally important role in the technology transfer process from the science base. They have developed considerable resources, and some have developed valuable products and services.

Factors Contributing to Success

Management Strengths and Strategy:

As Fildes (1990) points out, managing technological excellence is the key to survival and growth in DBFs. Key amongst the factors which have contributed to Celltech's achievements, has been its internal management. Celltech is renowned amongst those involved in the industrial, scientific and public policy biotechnology community for being a well-managed company. [4] That is not to say that its management has been without its critics. [5] However, the style of management adopted, with its emphasis on creating good systems for communication and engendering high motivation, has been enthusiastically received within the company. [6] Importantly, the company has managed to overcome many of the thresholds which constrain growth, such as the rapid establishment of credibility and making the transition from being 'entrepreneurially' to 'professionally' managed. It quickly established business and scientific credibility as a company making products (anti-interferons), rather than purely being a technology transfer organisation. It built upon this early credibility as each major contract has been signed and has developed a considerable reputation for its ability to manufacture to very high standards. The company successfully made the necessary transition from executive control being in the hands of one person, to it residing in a broader franchise of decision-makers. And the Board of directors, under the chairmanship of John Jackson, made the transition from being originally investor-centred and disunited, to a more unified body representing all those involved with the company.

Another factor positively affecting the company's development has been the long-term nature of much of its decision making. A recent report highlights how the absence of strategic skills are a major barrier to growth in small firms (ACOST, 1990), and Celltech's commitment to strategic management closely resembles that found amongst some of Europe's leading high technology small and medium-sized firms (Dodgson and Rothwell, 1989b). Its adaptive commercial strategy has necessarily been long-term oriented with a continuing strategy of high commitment to technology generation and acquisition and to human resource development. The commitment to long-term aims is most clearly revealed in the decision made to reinvest considerable resources back into the company in line with its pharmaceutical strategy once the first profits had been made, seven years after its formation. Perhaps another example of its foresight is the appointment of a Marketing Planning Manager (Pharmaceuticals), six years before its intended launch date for its first pharmaceutical product. Another aspect of its long-term human resource development strategy has been its efforts to develop the management skills of the scientists it has recruited. Scientists, often from academic backgrounds, have been encouraged to develop novel skills, for example in financial controls and the man-management skills Celltech's employment system requires.

The management task at Celltech has been very considerable. On the one hand it has involved the development of a management with high standards and business ethics. On the other it has involved the creation of practical operational systems and procedures. Perhaps one of the central management tasks has been building up long-term competences: in developing and fusing together the technology base and the knowledge base of its staff; and at the same time realising shorter-term commercial objectives. Almost from the start Celltech has had products to sell, and it has successfully developed a business in the bulk production of biotechnology products. Considering the financial constraints on the company's ambition, controlling the inevitable tensions between long-term and short-term pressures has been a major managerial task. The tension between the long and the short-term is one that is apparent throughout the company: from the project and time allocation of individual scientists, to senior management decisions about project selection, to Board level discussions about strategic objectives. The fact that Celltech currently has commercial activities and long-term opportunities may provide some partial indication that these tensions have been managed and controlled. That is not to say, of course, that these tensions have been resolved.

Throughout this book, emphasis has been placed on the management of learning, and we have seen how adept Celltech is at this. Elements of technological learning have been discussed, internal and collaborative. Facets of organisational learning have been described as the company grew and adapted its structures in line with requirements. A number of examples have been provided of the ways in which the company has learnt through failures, and the way it responded to lessons about, for example, the importance of patents, or the necessity to build up marketing skills. There is little doubt that Celltech has learnt some salutary lessons about the management problems involved in running joint ventures: not least in ensuring convergence in partners' management strategies, practices and expectations.

Particularly important has been the considerable management efforts put into human resource development. It is the employment system, and the incentives and motivation it provides, which creates the basis for the company's entire strategy, and those components within it related to learning and technological competitiveness.

Management has also to be assessed in the light of the way that markets for biotechnology products have been evolving. When the company embarked on its original strategy, there was a great deal of uncertainty about the products that biotechnology could provide. The company kept its options open and adapted its strategy according to opportunity. The pursuit of such a strategy, without immediately discernible commercial targets, is risky, but has produced rewards. For example, there was scepticism originally about the commercial viability of the company's decision to pursue two technologies: MAbs and rDNA. While this undoubtedly compounded the company's financial problems, the company has benefited from pursuing both. Its competences in both have facilitated the development of its antibody engineering skills, and its two targeted short-term products, CGRP and anti-TNF, derive respectively from rDNA and MAbs technology. Similarly, the company's decision to invest in manufacturing plant was not market-led. Director of Development, John Birch, describes the investment as 'in part, an article of faith'. Although the decision to invest in this plant was undertaken in a very business-like way, it was still a risky and speculative departure.

No company can operate for ten years without its failures and Celltech is no exception. In a fast moving environment mistakes are inevitable. One of the company's most obvious failures has been its inability to create successful joint ventures from the opportunities it had in diagnostics and microbiology. Throughout its history, Celltech has been criticised from one source or another for its perceived lack of market focus. The validity of this criticism is not easily assessed, [7] particularly given the general uncertainties facing DBFs. Certainly in the past, Celltech has on a number of occasions failed in its aims for particular projects due to a variety of reasons, including: lack of awareness about competitor activity; lack of appreciation of the scientific problems to be overcome; insufficient understanding of market requirements. And we have seen that there have been past disagreements amongst some of the shareholders about Celltech's commercial acumen. There is a belief amongst a number of past and present Celltech employees that the company, for a variety of reasons, was not as quick as it could have been in focusing its commercial targets in its early years. It was not until 1982/83 that the company recruited senior staff knowledgeable about marketing in the pharmaceutical industry. There remains a current of opinion outside of the company that identification of products with major commercial potential has not been one of Celltech's greatest strengths. [8] Such criticisms have, of course, to be put in the context of the company's commercial successes, shown most clearly by its steady growth in sales.

The absence of strict commercial focus in its early years, and the emphasis on a science and technology-led strategy, may paradoxically (from the point of view of those critical of the company's commercial acumen) eventually produce greater commercial returns to the company's investors. Director of Sales and Marketing, John Berriman, argues that if Celltech had selected a product, or retained the rights on one or more of its products, early on it would not have had the resources to spread its expertise. The company's technology would have become very specialised and it would not be able to capitalise on the possibilities provided by antibody engineering.

Luck:

Success in business depends on a number of factors, not least luck. Celltech has been bedevilled by problems, but it has also ridden its luck. This is perhaps most apparent in a number of senses concerning the development of its technology and in the products to emerge from its research. Few people foresaw the convergence of rDNA and MAb science in antibody engineering, so Celltech's decision to pursue both areas of science, providing it with the considerable potentials antibody engineering offers, has to be put down to serendipity (it did, however, quickly pick up opportunities this provided). Similarly, the potential products developed from Celltech's own-product research were not specifically devised to target critical care: markets more suitable to Celltech, but they fortuitously emerged so. However, fortune favours the prepared mind. And Celltech would not have been in any position to take advantage of these developments had it not had its technological expertise and organisational structures in place.

Constraining Factors

Technological Uncertainty:

In Chapter 2 the concept of 'technological discontinuity' was introduced. It was argued that the time factor was very important in deciding the outcomes of radical technological change, and in establishing how 'discontinuous' it was to be for established firms (and thereby what opportunities exist for new entrants). Biotechnology is rapidly developing and has very considerable potential. However, the processes of adjustment to the technology continues. There has been an underestimation of the time necessary for important scientific and technological discoveries with the potential to produce commercial and widely diffused new products to realise that potential. Semiconductors – a tangible product – invented in the late-1940s, took at least 25 years to develop into the range of applications we know today. The two major biotechnology breakthroughs of rDNA and MAbs – process inventions with no immediate products – only occurred in the mid-1970s. It is unrealistic to have expected products to have emerged, and to have been widely diffused, in such a short period.

Commercial and Regulatory Uncertainties:

Commercial uncertainties abound in biotechnology, for example, about the extent of competition, the efficacy of therapeutics and their costs, and the ability to achieve patent protection. These uncertainties have manifested themselves in the wildly fluctuating attitudes of analysts and investors towards biotechnology (see Teitelman, 1989). Regulatory uncertainties have plagued the whole development of biotechnology, aided by the early deep suspicions amongst certain powerful elements about the morality of manipulating genes (see Hall, 1987). Doubts remain on what the European Community's position will be on regulations. Patent disputes continue to occur and some, such as Genentech versus Wellcome over t-PA, take many years to resolve, and add considerably to the uncertainty surrounding biotechnology.

Divergent Expectations:

Teitelman (1989) highlights how broad the range of expectations is from biotechnology from scientists, managers and venture capitalists. These differences have clearly been seen at Celltech, and had an important effect. A number of the disagreements on Celltech's Board have been described, most particularly the problems related to divergent views about a number of investments. These essentially concerned differences in opinion about whether the company should increase its investment in line with opportunity before it had returned a profit. From its beginning Celltech did not have a group of shareholders united in purpose. Celltech emerged from a public policy for science and technology disrupted by the Conservative election victory in 1979. It inherited a legacy from this time – a heterogenous group of shareholders, with differing expectations – which can be argued to have occasionally disrupted its development as a company with a long-term, science-based strategy. Certainly arguments with TDC and BIL took up a great deal of Celltech management's time and energies. Whether these organisation's insistence on closer controls over management decisions and greater commercial prioritisation amended the behaviour of Celltech's managers, or were indeed necessary, is an open question. It is unlikely, however, that the experienced representatives of the other shareholders, and the chairman, would tolerate inefficient management.

The problems the company faced in having a divergent group of shareholders derived from 1979/80 when the plans for the company were being developed. Although Celltech was the brainchild of two public-sector organisations – the NEB and MRC – political dictum did not allow it to be public-sector owned. Private sector backing was required, and the combination of severe time pressures placed on the NEB organising team by numerous bureaucratic delays, [9] and the limited enthusiasm in the City for investing in biotechnology and in NEB initiatives, ensured that Celltech began its life with its uneasy group of shareholders. [10]

To add to its insecurity, Celltech's major shareholder, the NEB, had made assurances that it would attempt to divest its holding within three years. Nevertheless, the NEB/BTG's financial support was critical to the company and arguably therefore to biotechnology in Britain. Its policy towards Celltech, however, changed over time. After it had made its initial investment, the NEB came under political and economic pressures to reduce its shareholdings as rapidly as possible. And senior NEB/BTG officials were trenchant in their view that Celltech should become a short-term, commercially-oriented company. 'National interest' considerations came more to the fore when BTG, relieved of the pressure of being the dominant shareholder and more secure in the knowledge that it would enjoy financial returns, began to make its decisions about its shareholdings in line with the needs of the company. There is little doubt, however, that the NEB/BTG's occasionally reluctant, but continuing, support was instrumental for Celltech's survival and growth.

Response from Large Firms:

Pisano (1989) compares the development of semiconductors in smaller firms in the USA, where defence and aerospace contracts underwrote much of the cost and risk of R&D, with the development of biotechnology where such support was not forthcoming. This leads him to emphasise the importance for DBFs of large-scale and long-term linkages with large companies, a common occurrence in the USA.

Celltech was not well received by the majority of large UK pharmaceutical firms. According to some, they reacted negatively to its agreement with the MRC. One senior member of Celltech's staff recalls large firms' attitudes at the time responding to an image of Celltech being brash and immodest in its ambitions. [11] The company did not negotiate a significant contract with a major UK pharmaceutical firm until May 1988. This has perhaps been one of the most important factors constraining the growth of the company. [12] As we have seen, many of the world's largest US, Japanese and European pharmaceutical firms have worked with Celltech, and its two current 'corporate' shareholders are Japanese and American firms. A number of these European, Japanese and American companies have in the past attempted to acquire significant shareholdings in the company. Virtually every collaborative contract Celltech has initiated has first unsuccessfully been offered to a British firm. British pharmaceutical companies have been offered the opportunity to buy stakeholds (without Board seats) in Celltech without success, and none purchased significant numbers of shares during Celltech's latest placing. It is curious that British pharmaceutical companies, amongst the most innovative and successful in the world, have been impervious to the attractions that Celltech provides to their competitors.

Unsupportive Public Policy Environment

Celltech's lack of recognition and support by large companies has been matched by the government. The company has not been supported by the government as a national asset undertaking leading research with potential benefits to the economy. [13] Throughout its history the company has operated in public policy circumstances where – unlike in many of Britain's competitors – intervention to assist long-term oriented, science-based companies has not been forthcoming. Celltech began its life in very difficult political circumstances. It was conceived under a Labour Government and born under a Conservative one. Its formation was supported by politicians and policy-makers not as a worthwhile and exciting venture with high potential, but as a reaction to the rigidities of a system for commercialising British science which had resulted in lack of recognition of an opportunity to patent a discovery of worldwide significance. [14] It is argued that the government has never had a coherent policy for biotechnology (Sharp, 1989). While the company has benefited directly from its links with the NEB/BTG and the MRC, and to a limited extent from links with the SERC and the DTI, [15] Celltech has not been generously endowed with government assistance. The government's present policy towards innovation has been to promote 'pre-competitive R&D' and avoid 'near market' intervention. We have seen, however, that only 2 per cent of Celltech's R&D efforts involve activities promoted by the government. [16]

Support from Core Financial Backers:

All DBFs have had to rely on long-term financial investments as few have been in the position to create sufficient income to support the necessary levels of R&D. Unlike the IT industry where small firms often had products immediately to sell, and thereby enabled self-funding for R&D, DBFs have had to rely on venture capital, lucrative R&D contracts, and shareholding deals with large firms. In the latter two cases there has been a limited number of firms prepared to enter such arrangements. We have seen how Celltech embarked on research programmes in two technologies on the basis of funds provided for one, and the problems its management had in its early attempts to increase available funds. The company has not had the opportunities that its US competitors have had in access to technology-based venture capital investments, nor a stock market which, despite its wildly fluctuating attitudes to biotechnology, has on occasion been extremely receptive to it. Some US biotechnology companies, such as Genentech and Cetus, had raised more finance by 1980 than Celltech has managed to do by 1989. [17]

Obviously very little of what Celltech has achieved would have been possible without shareholders prepared to take the long-term view. These include three which have invested for a lengthy period in Celltech: British and Commonwealth Shipping, the Prudential and Midland Bank (in its various guises). These shareholders have done something which City investors are often accused of not doing: investing long-term in smaller science and technology-based companies.

The paucity of investors in Britain such as those Celltech has had over the past ten years has been thrown into stark relief by the problems it currently faces in finding a purchaser for British and Commonwealth's share. These shares are for sale as British and Commonwealth – a financial services conglomerate – is in the process of being wound up, having succumbed to the aquisition mania of many City institutions and highly unwisely taking over a US company whose debts have brought the whole group to collapse. This situation throws into sharp relief the precarionsness of firms like Celltech operating with such limited financial options. Its efforts to build a 'learning organisation', and all its structures and systems, depends crucially on long-term financial security. Without this, it cannot continue its existing strategy.

Future Risk

The acquisition of Genentech by Hoffman-La Roche, stimulated by Genentech's realisation that it independently would lack the enormous resources needed to develop a stream of biotechnology products, must raise questions about the ability of DBFs to become major pharmaceutical companies. Attracting sufficient levels of finance provides the greatest problem facing DBFs. [18] The enormous costs involved in the lengthy process of getting drugs to market do not, of course, end with the $100 million or so for getting approval. Marketing costs can be huge. To cover the USA a big drug company would normally need 2000 marketing personnel (Financial Times, 7 June 1989).

Celltech's pharmaceutical strategy provides its primary element of future risk. For the strategy to succeed Celltech requires long-term, and additional capital. Although it is expected within Celltech that it will be able to launch its drugs very much more cheaply than $100 million, if one or both of CGRP or anti-TNF is successful, then the company will nevertheless have to seek extra funding. The question is: how risky is the strategy? The problems to be overcome in realising this strategy are similar to the ones Celltech has faced throughout its existence, although at a much higher level, of course: are the correct products targeted (i.e. are they scientifically feasible, protectable, and are markets large enough); does the company have the right skills to realise its ambitions; and, crucially, does it have sufficient financial resources?.

The company is confident about the indications for its two prioritised drugs. Such confidence is necessary given how very few new products reach the international drug market each year. It is estimated that only 52 new products were launched on world markets in 1988. In 1986 only 44 new products emerged, at the same time as some 6,100 compounds were under development (Financial Times, 7 June 1989). The way that the two products focus on critical care, requiring a comparatively small marketing effort operates in the company's favour. Genentech's recent victory over Wellcome concerning the patenting of tPA shows that patents can be protected. Certainly on the question of having the right skills, Celltech has shown itself to be extremely adept at integrating new skills, either within the company or in collaboration.

On the question of raising funds, the difficulties of the situation are compounded by the ownership problem. The company has delayed its planned public flotation until its plans for its products have become more concrete. If the efficacy of these products is established, then Celltech's greatest need will be, as so often in the past, patient and sufficient financing. Funding is essential not only to get the drugs to market, but to ensure that present and future generations of scientists are sufficiently attracted to work on Celltech's R&D programmes, and to create a pipeline of new products. The question remains on how Celltech can attract shareholders supportive of this strategy.

The Central Importance of Learning

Fast learning is a critical factor affecting the competitiveness of DBFs. We have seen the centrality of learning within Celltech, and the processes by which it occurs. It has been the explicit intention of the Celltech's learning-based strategy to build its technology and knowledge bases. The technology base in Celltech, and its abilities to supplement that base through networking, are the creation and tools of its workforce. In the company the maxim that the strength of an organisation lies in its people is particularly apparent. Very considerable managerial efforts have gone into fostering a well organised, committed and creative workforce. Building this workforce, with its extensive knowledge base, is a long-term endeavour. The employment system is a crucial element in the company's strategy. It provides the means by which technological know-how is brought into, generated, diffused and retained within the company. The extent of its learning abilities have enabled it to pursue its adaptive business strategy, a strategy that has proven particularly advantageous in circumstances of considerable technological and competitive uncertainty.

The strategic approach to learning, and the actual ways that the company has gone about learning, may hold some lessons for all those firms confronted by rapid technological change. Celltech is, of course, comparatively new and small. It could be argued that this provides very particular circumstances and is not applicable to established firms in any broader sense. Certainly newer firms may have significant differences. There is no requirement for 'unlearning', for example, and any learning through experience may be constrained by the lack of time the firm has been operating. However, the point is that many of the lessons of newness of firm also apply to newness of technology. And hence the emphasis on, and nature of, learning described above may well be relevant to a broader population of firms, and may include those attempting to work with, and capitalise upon, a radical new technology.

At the time of writing Celltech's future is in the balance. This is also true for a number of DBFs with similar ambitions. Despite all its efforts and not inconsiderable achievements Celltech is suffering the vagaries and shortcomings of the financial markets. Will it go the way that so many of the world's leading DBFs are going, and be bought-out by a large firm? [19] Or will it continue its independent path with additional corporate investors? Only the future will tell. Pierre Vergniaud, describing political events in France two hundred years ago, told us that revolutions devour their young. Will the biotechnology 'revolution' do the same? One thing is clear. The future for DBFs will depend crucially on that factor which has enabled them to contribute so extensively to the development of biotechnology in industry: their comparative advantage in their ability quickly to learn about, and effectively use, radically new and rapidly changing technology. It is this which makes DBFs financially attractive propositions. It has been the aim of this book to analyse how this learning takes place and the organisational and employment system factors which encourage fast learning. The strategic management of learning has been shown to be a critically important facet of the task of dealing with and benefiting from turbulent technological change. Building organisations to maximise learning capabilities is a long-term process, requiring considerable commitment from a firm's management and shareholders. As it was in the past, so it will be in the future.

Footnotes

[1] Market capitalisation of a private company is not too meaningful a measure and is in any case difficult to assess. However, by multiplying issued shares by their modal price over the last year, a rough figure of £120 million is derived. It should be emphasised that this figure may not represent the actual value of the company. Teitelman (1989) records the considerable problems Wall Street has faced when attempting to assess the value of biotechnology companies. Some of the methods used have been eccentric, and few, according to Teitelman, have been particularly revealing.

[2] It is an open question how many scientists would have joined the brain drain had Celltech not existed. One suspects that, given the few available alternatives in the UK, this number might have been significant.

[3] For example: Dr Tim Harris left to become Head of Biotechnology at Glaxo; Dr Mike Boss and some colleagues left to start a spin-off company, Oros Instruments Ltd; and Dr Chris Hentschel left to run the MRC's Collaborative Centre.

[4] This is apparent in its systems such as in financial controls, but Celltech also has has the 'feel' of being professionally organised. Security guards are always vigilant; you

can always get through on the 'phone; staff are constantly aware of health and safety issues. To illustrate the latter observation; on one occasion an executive director and I were just leaving an R&D laboratory, and had forgotten to take our laboratory coats off (which are not allowed to leave the R&D area). Within seconds we were pulled up by a safety representative. The company has a full-time health and safety officer.

[5] Inevitably this has focused on the chief executive. Particular criticisms of him in the past, made by people outside the company, include that he is too open in his style of management, too long-term in his commercial thinking, and has, on occasion, shown excessive loyalty to employees. One other criticism is made of his style; and that is the robustness of his defence of the company.

[6] Much of this is derived from the commitment shown to it by Gerard Fairtlough, and the respect accorded to him. The employees I have interviewed, both past and present, hold the chief executive in considerable esteem. Many of the Board members I have spoken to pay tribute to the extraordinary talent Fairtlough has in building up the loyalty and commitment of those who work for him.

[7] Specific criticisms made by people outside the company have often been contradictory: 'it is too adventurous in its strategy'; 'it is too conservative'; 'it has been too scientifically-led', 'it hasn't spent enough on science'. Others have been obviously misinformed. Two people interviewed, for example, believed that the company has not targeted Europe sufficiently. Europe is, of course, a major target for its pharmaceutical strategy. More difficult to assess have been those criticisms of the company's commercial targeting. Past performance can be evaluated. Future outcomes, dependent on activities undertaken during the period of this study, are uncertain. As the author has limited knowledge of the likelihood of product developments succeeding, and of their potential market size, he has had to rely on the views of experts external to the company.

[8] See, for example, the remarks of an unidentified 'venture capital expert' in an article by Peter Marsh in Financial Times, 14 November 1989. However, in the same article a healthcare analyst at the stockbrokers Kleinwort Benson, Mr Ian White, stated that he was confident that Celltech would become a significant pharmaceutical company by the end of the century.

Celltech has minority shareholding in a number of small biotechnology companies, and it is ironic – given the criticisms from some – that Celltech feels that it in a position to offer such firms assistance not only with finance, but also with commercial direction.

[9] Delays caused by the resignation of the original NEB Board; Keith Joseph's insistence that the new incoming NEB Board prioritise electronics companies; and the tortuous negotiations between the NEB, NRDC and the Treasury. These delays compounded the fears that US companies were well ahead in their drug discovery programmes. In 1985 Gerard Fairtlough said that had Celltech been in the USA, it probably would have been off the ground 18 months earlier (Wall Street Journal, 9 August 1985).

[10] Its five shareholders represented: the public sector (the NEB); a commercial company (British and Commonwealth Shipping); a financial services company (the Prudential); a bank (the Midland); and a venture capital company (TDC). None had experience of investing in biotechnology companies.

[11] New biotechnology start-ups have a reputation for immodesty. See Julian Davies' review of Teitelman (1989), Nature, Vol.342, 2 November 1989.

[12] In 1984, Kenneth Baker, then Minister for Information Technology, was quoted as saying 'established companies in Britain ... must get alongside Celltech', and he commended the liaison between Boots and Celltech (Statement at Biotech Europe 84 PR, 15 May 1984).

[13] Many people involved in biotechnology in Britain believe that Celltech is a strategically important company to the nation. Dr Brian Richards, for example, says 'It is very important for the UK that Celltech succeeds'. It is reported that Celltech influenced the establishment of the state-aided French biotechnology company, Transgene (Russell, 1988). The company has provided a model which a future Labour Party government would wish to expand (see, J. Bray, Labour spokesman on science, Laboratory News, 371, 27 March, 1987).

[14] See Chapter 3, particularly the comments of Lord Joseph. Professor Shaw, a member of Celltech's Science Council from its formation, describes the company's creation as a 'breathless response to the failure to patent monoclonal antibodies'.

[15] Through joint 'club' activities, and also in the past Celltech has received some small but not insignificant DTI grants towards development of manufacturing processes.

[16] There is a feeling within the company that it is not favoured by the government. From a government which misses few opportunities to promote British successes, particularly smaller, high technology firms (such as Sinclair), this is perhaps surprising. Mrs Thatcher apparently shows an interest in Celltech, but she has declined invitations to visit, although she has visited a number of US biotechnology firms. Celltech does, of course, provide an example of a successful investment by the state, counter to the non-interventionist beliefs of the government.

[17] Celltech's financial problems have been compounded, of course, by its unfortunate timing of the 1987 share placing. British Bio-technology Ltd has managed to raise nearly half as much Celltech has done in nine years since 1986 (£33 million to £70 million).

[18] See, for example: 'The money guzzling genius of biotechnology', The Economist, 13 May 1989; and Raunch, L 'Surviving the Funding Gap', Bio/Technology, Vol.7, May 1989. The survey by Burrill (1989) of 480 US biotechnology companies found that 62 per cent of them needed major new financing by the end of 1990. The larger, longer established US DBFs are comparatively better off in this respect. It is reported that in 1988 Cetus enjoyed cash reserves of $240 million and Genentech, $180 million (Biotechnology Bulletin, Vol.6, No.12, January 1988).

[19] Burrill's (1989) survey showed that nearly one-half of surveyed biotechnology executives expected their companies to be acquired within the next ten years.

References

Ansoff, I (1982) 'Managing Discontinuous Strategic Change: the Learning-Action Approach' in Ansoff, I, Bosman, A and Storm, P Understanding and Managing Strategic Change, North Holland, Amsterdam.

Argyris, C and Schon, D (1978) Organizational Learning, Addison-Wesley, London.

Arrow, K (1962) 'The Economic Implications of Learning by Doing' Review of Economic Studies, June.

Bell, M and Scott-Kemmis,D (1990) 'The Mythology of Learning-by-Doing in World War II Airframe and Ship Production', Science Policy Research Unit, March.

Beynon, H (1973) Working for Ford, Penguin, London.

Beynon, H and Nichols T (1977) Living with Capitalism, RKP, London.

Burawoy, M (1979) Manufacturing Consent, University of Chicago Press, Chicago.

Burns, T and Stalker, G (1961) The Management of Innovation, Tavistock, London.

Burrill, G (1989) Biotech 90: Into the Next Decade, Ernst and Young, New York.

Calori, R (1990) 'Effective Strategies in Emerging Industries' in Loveridge, D and Pitt, M The Strategic Management of Technological Innovation, Wiley, Chichester.

Chandler, A (1962) Strategy and Structure, McGraw Hill, New York.

Chesnais, F (1988) 'Technical Co-operation Agreements Between Firms' STI Review.

Cyert, R and March, J (1963) A Behavioural Theory of the Firm, Prentice Hall, Englewood Cliffs.

Daly, P (1985) The Biotechnology Business: A Strategic Analysis, Pinter, London.

Dodgson, M (1989a) 'Strategies for Technological Learning: New Forms of Organisational Structure' Paper presented at APROS Conference on Organisations, Technologies and Culture, Australian National University, Canberra, December.

Dodgson, M (1990a) 'Celltech – the First Ten Years of a Biotechnology Company' Special Report, Science Policy Research Unit.

Dodgson, M (1990b) 'The Shock of the New: The Formation of Celltech and the System of Technology Transfer in Britain', Industry and Higher Education, June.

Dodgson, M (1989b) Technology Strategy and the Firm: Management and Public Policy, Longman, Harlow.

Dodgson, M and Rothwell, R (1989a) 'Financing Early-Stage Innovation in Small and Medium-Sized Firms' paper presented at the European Conference, Enterprise, Innovation and 1992, Nice, October 26-27.

Dodgson, M and Rothwell, R (1989b) 'Technology Strategies in Small and Medium Sized Firms' in Dodgson, M Technology Strategy and the Firm: Management and Public Policy, Longman, Harlow.

Dore, R (1973) British Factory – Japanese Factory, University of California Press, Berkeley.

Dosi, G (1982) 'Technological Paradigms and Technological Trajectories. A suggested Interpretation of the Determinants and Direction of Technological Change, Research Policy, Vol 3, No 3.

Dosi, G (1989) 'Sources, Procedures and Microeconomic Effects of Innovation', Journal of Economic Literature, Vol XXVI, September.

EVCA (European Venture Capital Association) (1989) EVCA Yearbook

Fairtlough, G (1986) 'Creative Compartments' London Business School Journal, Summer.

Fairtlough, G (1989) 'Systems Practice from the Start: Some Experiences in a Biotechnology Company', paper presented at the UK Systems Society, 26th May. To be published in Systems Practice (forthcoming).

Faulkner, W (1988) 'The British Biotechnology Industry', draft mimeo, University of Edinburgh.

Fildes, R (1990) 'Strategic Challenges in Commercializing Biotechnology' California Management Review, Spring.

Freeman, C (1982) The Economics of Industrial Innovation, Pinter, London.

Freeman, C and Perez, C (1988) 'Structural Crises of Adjustment: Business Cycles and Investment Behaviour', in Dosi, G et al Technical Change and Economic Theory, Pinter, London.

Georghiou, L, Metcalfe, S, Gibbons, M, Ray, T and Evans, J (1986) Post Innovation Performance, Macmillan, London.

Hacking, A (1986) Economic Aspects of Biotechnology, Cambridge University Press, Cambridge.

Hagedoorn, J and Schankenraad, J (1990) 'Inter-Firm Partnerships and Cooperative Strategies in Core Technologies', in Freeman, C and Soete, L New Explorations in the Economics of Technical Change, Pinter, London.

Hagedoorn, J and Schot, J (1989) 'Co-operation Between Companies and Technological Development', mimeo, MERIT, University of Maastricht, February.

Hambrick, D and Crozier, L (1985) 'Stumblers and Stars in the Management of Rapid Growth', Journal of Business Venturing, Vol 1.

Hall, S (1987) Invisible Frontiers: The Race to Synthesize a Human Gene, Sidgwick and Jackson, London.

Hamilton, W, Vila, J and Dibner, M (1990) 'Patterns of Strategic Choice in Emerging Firms: Positioning for Innovation in Biotechnology' California Management Review, Spring.

Harris, R and Mowery, D (1990) 'Strategies for Innovation: An Overview' California Management Review, Spring.

Hedberg, B (1981) 'How Organizations Learn and Unlearn' in Nystrom, P and Starbuck, W Handbook of Organizational Design: Vol 1, Oxford University Press, Oxford.

Horwitch, M (1986) Technology in the Modern Corporation: A Strategic Perspective, Pergamon, New York.

Laing and Cruickshank (1982) Biotechnology, London, March.

Loveridge, R and Pitt, M (1990) The Strategic Management of Technological Innovation, Wiley, Chichester.

Maidique, M and Zirger, B (1985) 'The New Product Learning Cycle' Research Policy, 14.

Metcalfe, S and Tiler, C (1989) 'Innovation and Strategic Management in SMEs; Post 1992', Paper presented at the European Conference, Enterprise, Innovation and 1992, Nice, October 26-27.

Morgan, K and Sayer, A (1989) Microcircuits of Capital, Polity Press, Cambridge.

Nelson, R and Winter, S (1982) An Evolutionary Theory of Economic Change, Belknap Press, Cambridge, Mass.

Oakey, R, Faulkner, W, Cooper, S and Walsh, V (1990) New Firms in the Biotechnology Industry, Pinter, London.

OECD (1989) Biotechnology: Economic and Wider Impacts, OECD, Paris.

OTA (1988) New Developments in Biotechnology: 4: US Investment in Biotechnology, Office of Technology Assessment, US Congress.

Orsenigo, L (1989) The Emergence of Biotechnology, Pinter, London.

Paine Webber (1989) Paine Webber's Financial Factbook for the Biotechnology Industry, Paine Webber, New York.

Patel, P and Pavitt, K (1988) 'The International Distribution and Determinants of Technological Activity' Oxford Review of Economic Policy, Vol 4, No 4.

Pavitt, K (1987) Comment on Tushman, M and Anderson, P 'Technological Discontinuities and Organizational Environments' in Pettigrew, A The Management of Strategic Change, Blackwell, Oxford.

Pisano, G (1989) 'The Governance of Innovation: Vertical Integration, joint Ventures, and Licensing in the Biotechnology Industry', mimeo, Harvard University, March.

Pisano, G, Shan, W and Teece, D (1988) 'Joint Ventures and Collaboration in the Biotechnology Industry', in Mowery, D International Collaborative Ventures in US Manufacturing, Ballinger, Cambridge, Mass.

Porter, M (1985) Competitive Advantage: Creating and Sustaining Superior Performance, Free Press, New York.

Pucik, (1988) 'Strategic Alliances, Organizational Learning, and Competitive Advantage: the HRM Agenda', Human Resource Management, Vol 27, No 1.

Rothwell, R (1983) 'Innovation and Firm Size: A Case for Dynamic Complementarity; Or, Is Small Really So Beautiful? Journal of General Management, Vol 8, No 3.

Rosenberg, N (1982) Inside the Black Box: Technology and Economics, Cambridge University Press, Cambridge.

Roy, D (1958) 'Banana Time: Job Satisfaction and Informal Interaction' in Salaman, G and Thompson, K (1973) People and Organizations, Longman, London.

Russell, M (1986) The Biotechnology Revolution, Wheatsheaf, Brighton.

Sahal, D (1982) Patterns of Technological Innovation, Addison Wesley, Reading, Mass.

Scott, M and Bruce, R (1987) 'Five Stages of Growth in Small Business', Long Range Planning, Vol 20, No 3.

Senker, J and Sharp, M (1988) 'The Biotechnology Directorate of the SERC', Science Policy Research Unit.

Senker, P and Brady, T (1989) 'Corporate Strategy: Skills, Education and Training', in Dodgson, M Technology Strategy and the Firm: Management and Public Policy, Longman, Harlow.

Sharp, M (1985) 'The New Biotechnology: European Government in Search of a Strategy', Sussex European Papers, No 15, Science Policy Research Unit, University of Sussex.

Sharp, M (1989) 'The Management and Coordination of Biotechnology in the UK 1980-88' Phil. Trans. R. Soc. Lond. B324, 509-523.

Sharp, M (1990) 'Technology Trajectories and Corporate Strategy in the Diffusion of Biotechnology' in Deico, E, Morneu, E and Vickery, G, Technology and Investment-Concial Issues of the 1990s, Pinter, London.

Snow, C and Miles, R (1983) 'The Role of Strategy in the Development of a General Theory of Organizations' in Lamb, R Advances in Strategy, JAI Press, New York.

Spinks, A (1980) 'Biotechnology: Report of a Joint Working Party' HMSO, London.

Teece, D (1987) The Competitive Challenge: Strategies for Industrial Innovation and Renewal, Ballinger, Cambridge, Mass.

Teitelman, R (1989) Gene Dreams: Wall Street, Academia and the Rise of Biotechnology, Basic Books, New York.

Tushman, M and Anderson, P (1987) 'Technological Discontinuities and Organizational Environments' in Pettigrew, A The Management of Strategic Change, Blackwell, Oxford.

Tushman, R, Virany, B and Romanelli, E (1986) 'Executive Succession, Strategic Reorientation, and Organisational Evolution', in Horwitch, M Technology in the Modern Corporation: A Strategic Perspective, Pergamon, New York.

Valentin, F (1989) 'When Companies Learn' Copenhagen School of Economics and Social Sciences.

Waite, R, Pearson, R, Varlaam, C and Pike, G (1988) Report for the Science and Engineering Council. Evaluating the Biotechnology Directorate's Training Awards Policies, Institute of Manpower Studies.

Walker, S and Parrish, J (1988) 'Innovation and Drug Development' in Walker, B and Walker, S Trends and Changes in Drug Research and Development, Kluwer, Lancaster.

Yanchinski, S (1980) 'Microbes at Work: is the Price Too High?', New Scientist, 3rd July.

Yarrow, D (1988) 'US Biotechnology in 1988 – A Review of Current Trends', Department of Trade and Industry, London, 1989.

Appendix 1
Interview Respondents

Apart from employees of the company, the following people were kind enough to allow me to interview for my study of Celltech.

Mr Ron Artus
Professor John Ashworth
Dr Sydney Brenner
Dr Jim Cain
Mr Mark Cantley *
Mr Wynne Denman
Mr Robert Fildes*
Mr David Fishlock*
Sir James Gowans
Dr Tim Harris*
Mr Roger Hay
Dr Chris Hentschel
Lord Joseph
Mr Charles Kirkman
Sir Arthur Knight
Mr Bill Matthews
Mr Hugh Perrott
Dr Brian Richards*
Mr Martin Sandford
Dr Derek Schafer
Professor William Shaw
Mr Karl Simpson*
Sir Michael Stoker
Dr Caroline Vaughan
Professor Bob Williamson*

* telephone interviews

Appendix 2
Members of Celltech's Board of Directors 1980-1989

1981	1982	1983
HW Denman	HW Denman	HW Denman
GH Fairtlough*	GH Fairtlough*	GH Fairtlough*
RE Artus	RE Artus	RE Artus
GA Gilhespy	J Cox	J Cox
JL Gowans	Sir JL Gowans	Sir JL Gowans
JBH Jackson	JBH Jackson	JBH Jackson
FA Russell	FA Russell	Sir F Tombs
tSir M Stoker	Sir M Stoker	Sir M Stoker
	Sir W Henderson	Sir W Henderson
	Sir D Phillips	Sir D Phillips
	DE Seymour	DE Seymour
		NH Carey*
		DW Gration*

1984	1985	1986
HW Denman	HW Denman	HW Denman
GH Fairtlough*	GH Fairtlough*	GH Fairtlough*
RE Artus	RE Artus	RE Artus
J Cox	J Cox	J Cox
Sir JL Gowans	Sir JL Gowans	Sir JL Gowans
JBH Jackson	JBH Jackson	JBH Jackson
Sir F Tombs	Sir F Tombs	Sir M Wood
Sir M Stoker	Sir M Stoker	
Sir D Phillips	Sir D Phillips	Sir D Phillips
DE Seymour	DE Seymour	DE Seymour
NH Carey*	NH Carey*	NH Carey*
DW Gration*	DW Gration*	DW Gration*

1987	1988	1989
HW Denman	P Myners	P Myners
GH Fairtlough*	GH Fairtlough*	GH Fairtlough*
RE Artus	RE Artus	RE Artus
J Cox		JE Berriman*
JBH Jackson	JBH Jackson	JBH Jackson
Sir M Wood	Sir M Wood	Sir M Wood
Sir D Phillips	Sir D Phillips	Sir D Phillips
DE Seymour	DE Seymour	DE Seymour
NH Carey*	NH Carey*	NH Carey*
DW Gration*	DW Gration*	DW Gration*
JG Huckle*	JG Huckle*	JG Huckle*
		JR Birch*

* Executive Director

Appendix 3
The History of Shareholdings in Celltech

Major share transactions and share prices are shown. Significant changes in percentage equity holdings (over 10 per cent) are shown, as are current major shareholdings.

November 1980: the founding shareholders committed £12 million in fully paid shares of par value of £1.00.

Shareholdings (% of equity):

National Enterprise Board	44%
Prudential Assurance	14%
Midland Bank	14%
British and Commonwealth Shipping	14%
Technical Development Capital (3i)	14%

Share price £1.00

August 1983: Celltech issues 3.6 million new ordinary £1 shares at £1.75 to raise £6.1 million. BTG sells 400,000 shares.

Shareholdings (% of equity):

British Technology Group	28%
British and Commonwealth Shipping	16%
Prudential Assurance	15%
Midland Bank	12%
Biotechnology Investments Limited	12%
Others*	17%

* includes: TDC, Royal Life Insurance, Prudential Staff Pension Scheme, Baring Brothers and Gartmore Investment Management.

Share price: £1.75

March 1985: Sale of existing shares to institutions and other investors. BTG sells 2 million shares.

Share price: £2.50

April 1986: Sale of existing shares to institutions and other investors. BTG sells remaining shares.

Share price: £3.00

July 1986: Celltech issues 1.57 million new shares which raises £4.9 million.
Share price: £3.00 – £3.20 (price dependent on various rights)
May 1987: Celltech issues 0.9 million new shares which raises £4.0 million.
Share price: £4.50

August 1987: Sale of existing shares between existing shareholders.
Shareholdings:

British and Commonwealth Shipping	36.5%
Prudential Assurance	13.9%
Midland Bank	12.2%
Gartmore Investment Management	6.8%
Baring Brothers managed funds	3.5%
Scottish Widows Fund	3.4%
Royal Life Assurance	3.0%
Sumitomo Corporation	2.4%
Johnson and Johnson Development Corporation	2.4%
Employees, Directors	3.5%
Various Investment/Life Assurance Companies*	12.4%

* Includes: Scottish American Investment, Prudential Mutual Life, Prudential Nominees, Norwich Union, Robert Fleming funds.
Share price: £5.17.5

December 1987: Placing of shares which raised £43 million.
Share price: £4.00

October 1989: Major shareholders:

British and Commonwealth Shipping	36.3%
+ Managed funds	2.8%
Prudential Assurance	13.3%
Midland Montague	7.2%
Gartmore Investment Management	6.8%
Sumitomo Corporation	2.4%
+ Investment funds	2.5%
Johnson and Johnson Development Corporation	1.5%
Others	27.2%

Share price: £4.00 to £4.20

Index

Bei Fragen zur Produktsicherheit wenden Sie sich bitte an:
If you have any questions regarding product safety,
please contact:

Walter de Gruyter GmbH
Genthiner Straße 13
10785 Berlin
productsafety@degruyterbrill.com